L. F

D0025082

APPROXIMATIONS FOR DIGITAL COMPUTERS

APPROXIMATIONS
FOR
DIGITAL COMPUTERS

BY CECIL HASTINGS, JR.

ASSISTED BY
JEANNE T. HAYWARD
JAMES P. WONG, JR.

PRINCETON, NEW JERSEY

PRINCETON UNIVERSITY PRESS

1955

Copyright © 1955, by The RAND Corporation
Published by Princeton University Press
All Rights Reserved

Second Printing 1957
Third Printing 1959
Fourth Printing 1962
Fifth Printing 1966

Printed in the United States of America

510.78
H357a
120114

Preface

THIS monograph deals with the subject of best approximation in the sense of Chebyshev as applied to the problem of making univariate functional data available to the high-speed digital computing machine. Our investigation is of a numerical and empirical nature.

Part I of this book serves as an introduction to the collection of approximations given in Part II. This exposition started its life as a film strip prepared for presentation to a local meeting of the Digital Computers Association in 1953. Much enlarged and considerably revised, the film strip is presented here with a running commentary under each frame.

Part II contains the "Approximations for Digital Computers," formerly issued as a cumulative publication of loose sheets and made available to numerical analysts upon request. Each sheet of the seventy-odd issued in this series contains an approximation of a useful or illustrative nature presented with a carefully drawn error curve.

The work presented in this volume was undertaken at The RAND Corporation in connection with its program of research for the United States Air Force.

Many acknowledgments are in order. Mr. J. D. Williams and other colleagues at RAND encouraged me to undertake this project. My assistants, Mrs. Jeanne T. Hayward and Mr. J. P. Wong, deserve special thanks for their help. Many members of RAND's Numerical Analysis Section assisted in various portions of the over-all task. My secretary, Mrs. Vada M. Baldwin, rendered constant and valuable service during the years in which the monograph was in preparation. Dr. Richard Bellman read and criticized the final manuscript. Special credit for the very accurate error-curve drawings in Part II is owed to Mr. I. G. Margadonna and to Mrs. Theresa Halverson of RAND's Publications Division.

CECIL HASTINGS, JR.

SANTA MONICA, CALIFORNIA
November, 1954

v

Contents

PART I

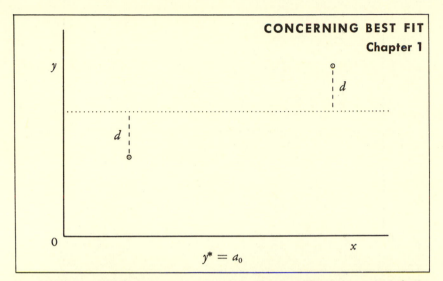

(1) *Minimum Absolute Error*: In the above drawing, the two vertical deviations from line to point have a common magnitude, d. We shall say that the dotted line represents a constant of best fit to the two points given.

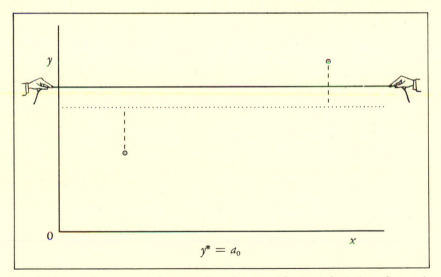

(2) We do so because any larger value of a_0 would give a deviation of magnitude greater than d for the point on the left.

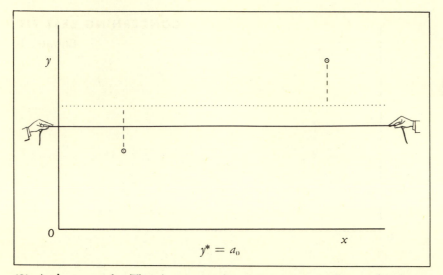

$$y^* = a_0$$

(3) And conversely. That is, we require our parameter values to be chosen so as to make the magnitude of the greatest deviation as small as possible. In the present instance, our one parameter is a_0.

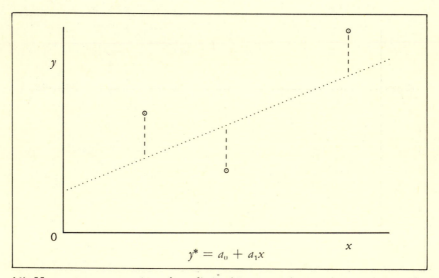

$$y^* = a_0 + a_1 x$$

(4) How can we recognize a best fit in this sense when we see one? Consider another example. In this frame there are three deviations of equal magnitude and strictly alternating sign. We claim that the dotted line is a line of best fit to the three points given.

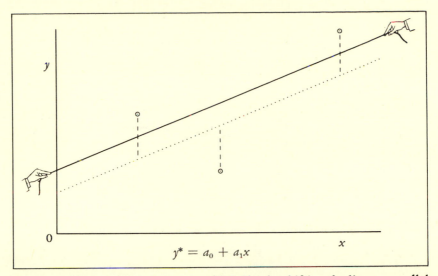

$$y^* = a_0 + a_1x$$

(5) Does this statement seem reasonable? Clearly, shifting the line up parallel to its old position will increase the middle deviation, and hence no improvement can result from such a change.

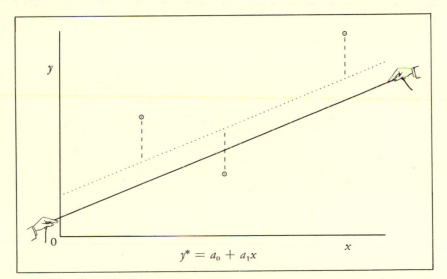

$$y^* = a_0 + a_1x$$

(6) And conversely. Thus, if an improved position is to be found, it must intersect the original position in some one of its points.

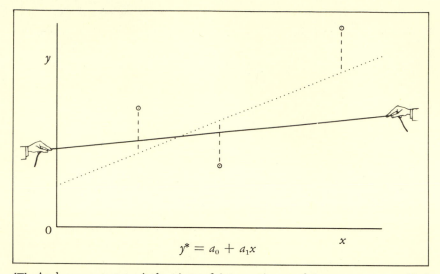

$$y^* = a_0 + a_1 x$$

(7) And so we test typical points of intersection, and in every case we find that a rotation in either sense increases the magnitude of at least one of the deviations. The dotted line is thus the line of best fit, as originally claimed.

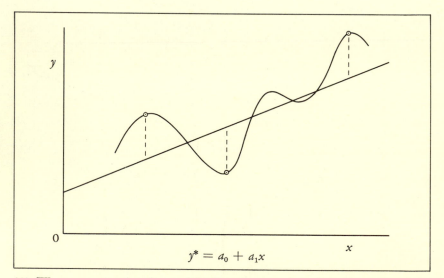

$$y^* = a_0 + a_1 x$$

(8) We now recognize the straight line above to be a best linear fit to the segment of curve pictured. *This is because no better fit can be made to the three encircled points on the curve.* An error curve for the above situation appears in the next frame.

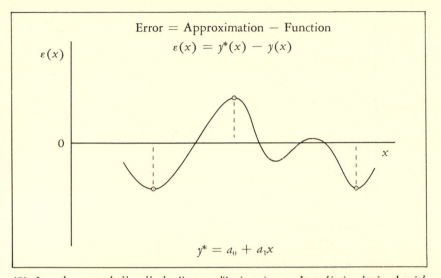

Error = Approximation − Function

$$\varepsilon(x) = y^*(x) - y(x)$$

$$y^* = a_0 + a_1 x$$

(9) *In what we shall call the "normal" situation, a best fit is obtained with a parametric form involving n parameters (two, in the case pictured) when n + 1 equal greatest deviations (three, in the case pictured) are obtained that alternate in sign.*

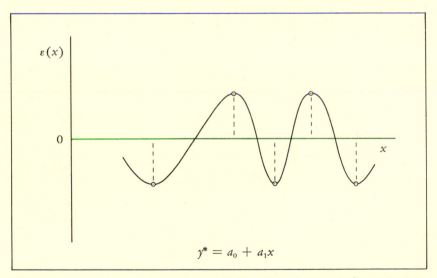

$$y^* = a_0 + a_1 x$$

(10) In rare instances of the "normal" situation there may be extraneous points at which the greatest error is achieved. In such a situation, the $n + 1$ points with the required property are not unique.

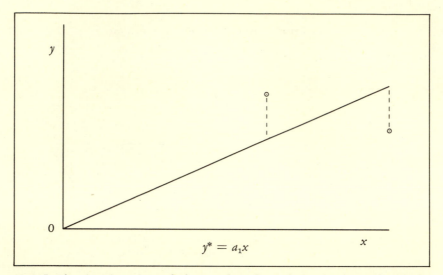

$$y^* = a_1x$$

(11) In the situation pictured above, a best fit is obviously obtained with the indicated form when there are two equal deviations of opposite sign. This is the normal situation: there is one parameter, and there are two deviations that alternate in sign.

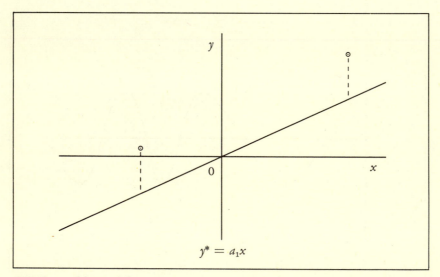

$$y^* = a_1x$$

(12) It would simplify matters greatly if all best-fit situations were "normal," but such is not the case. Thus, in the present instance, a best fit is obviously obtained with the indicated parametric form when the deviations are equal and of like sign.

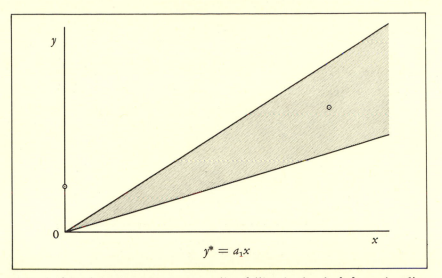

$$y^* = a_1x$$

(13) And in the present instance any line falling in the shaded area is a line of best fit. If this statement appears confusing, recall our definition of best fit, which merely states that the greatest deviation is to be made a minimum.

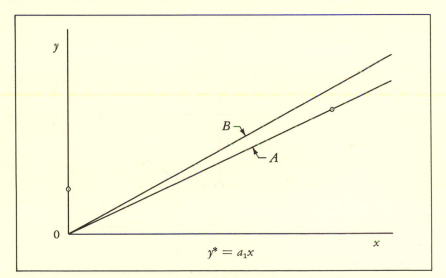

$$y^* = a_1x$$

(14) Thus, in accordance with our definition, line *A* is no better a fit to the two points indicated than is line *B*. In either case, the greatest deviation, the one at $x = 0$, is exactly the same and can be made no smaller.

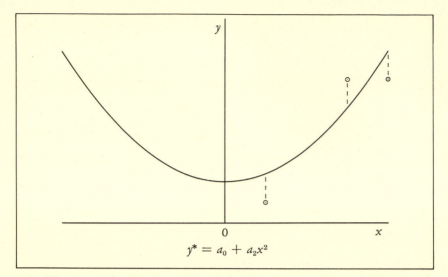

$$y^* = a_0 + a_2x^2$$

(15) Now, consider the situation pictured above. We obviously have a best fit, as the change of variable $\zeta = x^2$ for $x \geq 0$ transforms our problem into one of linear approximation to three points.

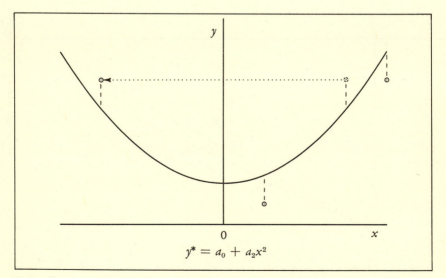

$$y^* = a_0 + a_2x^2$$

(16) If we now reflect the middle point in the y-axis, our curve of best fit must remain unchanged, as each instance of the form considered is symmetric about the y-axis. Our best-fit deviations no longer alternate in sign!

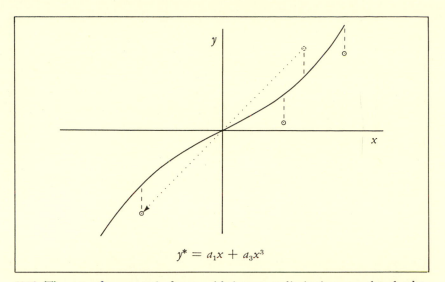

$$y^* = a_1 x + a_3 x^3$$

(17) The use of parametric forms with improper limitations can thus lead to "abnormal" situations of best fit. In particular, polynomial forms with uncalled-for gaps (missing terms of lower order) may be expected to give trouble if the interval of approximation includes the origin.

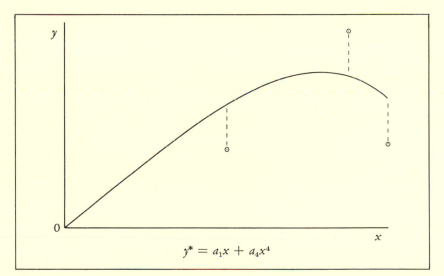

$$y^* = a_1 x + a_4 x^4$$

(18) However, the "normal" situation will always obtain in using a polynomial form with gaps if the data to be fitted have x-coordinates that are distinct and all of one sign. A proof of this statement and of several statements to follow is indicated in the problem section at the end of this chapter.

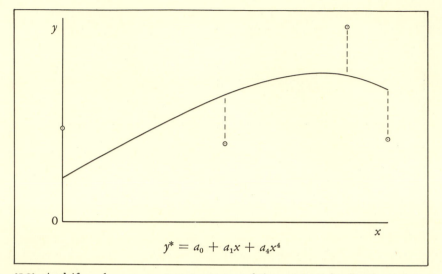

$$y^* = a_0 + a_1x + a_4x^4$$

(19) And if we have a constant term, one of the points to be fitted may even have a zero x-coordinate, and the "normal" situation will still obtain.

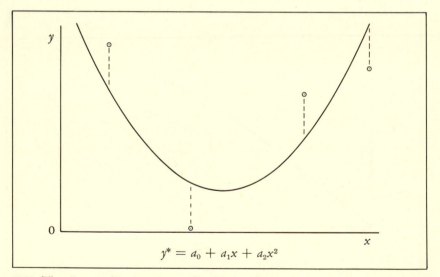

$$y^* = a_0 + a_1x + a_2x^2$$

(20) The "normal" situation will always obtain for the polynomial form without gaps if the x-coordinates of the data to be fitted are distinct. The x-coordinates of the data to be fitted need not be all of one sign.

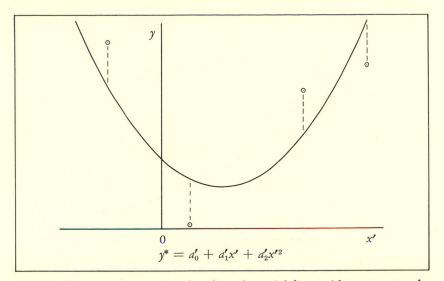

$$y^* = a_0' + a_1'x' + a_2'x'^2$$

(21) In this connection, notice that the polynomial form without gaps may be subjected to the transformation $x' = x + a$ without undergoing a change of form, in the sense that new types of terms in x' are added that do not appear in terms of x in the original form.

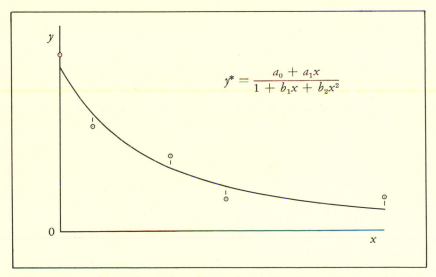

$$y^* = \frac{a_0 + a_1 x}{1 + b_1 x + b_2 x^2}$$

(22) Once we depart from the subject of simple polynomial approximation, there appears to be very little friendly theory to guide us. Considerable experience, however, leads us to expect the "normal" situation of best fit to obtain in all reasonable circumstances.

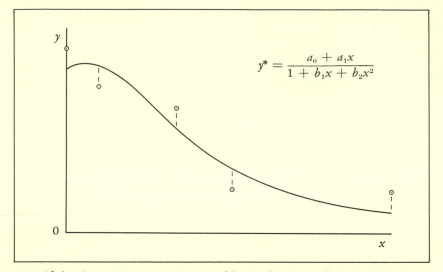

$$y^* = \frac{a_0 + a_1 x}{1 + b_1 x + b_2 x^2}$$

(23) If the circumstances are unreasonable, trouble generally arises in attempting to carry out the fitting process. In the above instance, one could run into considerable difficulty in attempting to find the continuous fit shown.

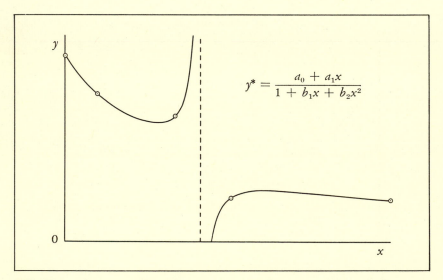

$$y^* = \frac{a_0 + a_1 x}{1 + b_1 x + b_2 x^2}$$

(24) This is because a discontinuous instance of the same form actually passes through the five points given.

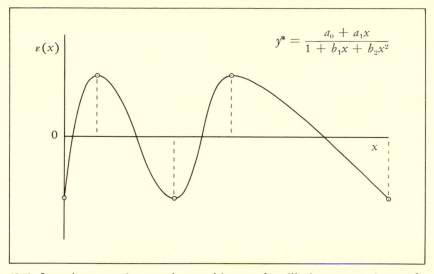

$$y^* = \frac{a_0 + a_1 x}{1 + b_1 x + b_2 x^2}$$

(25) Sometimes we give up when nothing we do will give us a continuous fit of "normal" appearance; but when we do obtain the expected situation, we feel quite certain that our fit is best.

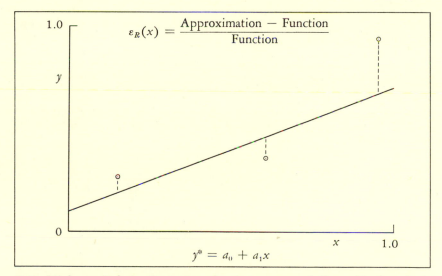

$$\varepsilon_R(x) = \frac{\text{Approximation} - \text{Function}}{\text{Function}}$$

$$y^* = a_0 + a_1 x$$

(26) *Minimum Relative Error*: In a few instances we shall consider best fit in the sense that the greatest relative error of approximation is made a minimum. Such cases will be clearly marked.

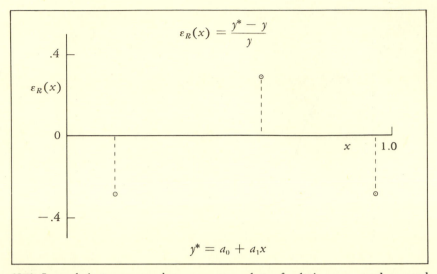

(27) In such instances we draw curves or plots of relative error and proceed to level the greatest deviations just as we would in the case of absolute error.

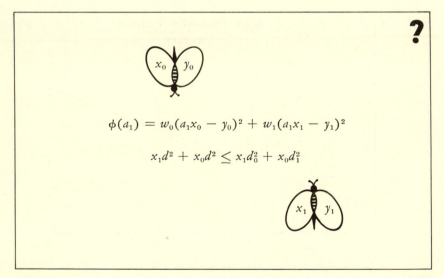

(28) We now indicate a bit of approximation theory which the reader may work out for himself. While the statement of Frame (11) is obvious, employ the above hints to yield an alternative proof of the fact.

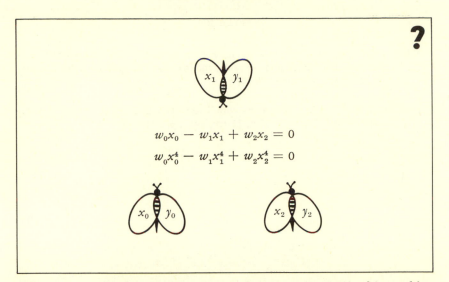

$$w_0 x_0 - w_1 x_1 + w_2 x_2 = 0$$
$$w_0 x_0^4 - w_1 x_1^4 + w_2 x_2^4 = 0$$

(29) With the further hint given here and the experience gained in working the previous problem, prove the statement of Frame (18) for the case illustrated there.

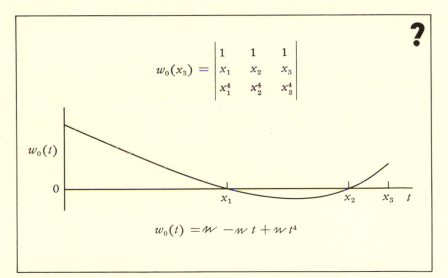

$$w_0(x_3) = \begin{vmatrix} 1 & 1 & 1 \\ x_1 & x_2 & x_3 \\ x_1^4 & x_2^4 & x_3^4 \end{vmatrix}$$

$$w_0(t) = w - w\,t + w\,t^4$$

(30) And with the further hint given here, prove the statement of Frame (19) for the case illustrated. If your work has been done as intended, it should be a simple matter to generalize the result considerably. For aid and further enlightenment, try the following references.

References

1. HASTINGS, CECIL, JR., "Rational Approximation in High-speed Computing," Proceedings of the Computation Seminar, International Business Machines Corporation, December, 1949, pp. 57–61.

2. SELFRIDGE, R. G., "Approximations with Least Maximum Error," *Pacific Journal of Mathematics*, Vol. 3, No. 1, March, 1953, pp. 247–255.

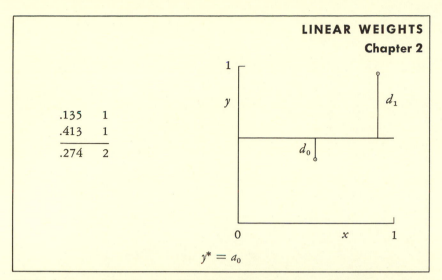

(1) *Prologue to "Concerning Weights"*: The deviations pictured above have magnitudes d_i and signs appropriate to that of best fit. Is it clear that we can determine d by taking a simple average of the two d_i?

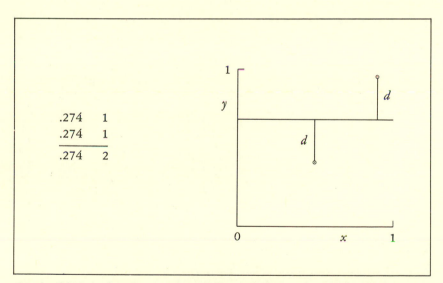

(2) As this is in fact the case, we shall say that $d_0 + d_1 = 2d$ is the invariant associated with the fitting of a constant to two points. With equivalent meaning, we shall say that $1:1$ weights level the deviations.

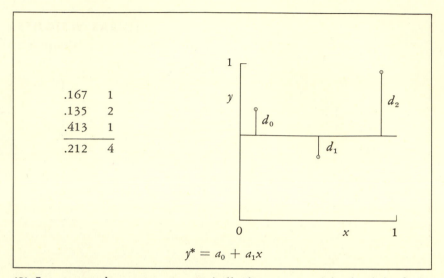

(3) Let us now demonstrate geometrically that $1:2:1$ weights level the deviations pertinent to a straight line and three equally spaced points.

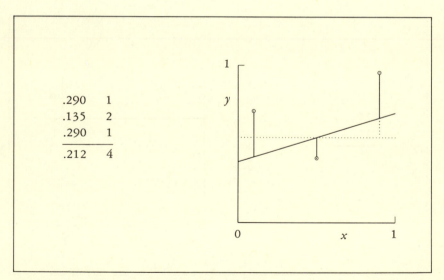

(4) Starting with the line in a horizontal position, we leave the middle deviation fixed and rotate the line a bit. As the two triangles to be seen are identical, it is obvious that $d_0 + d_2$ remains constant.

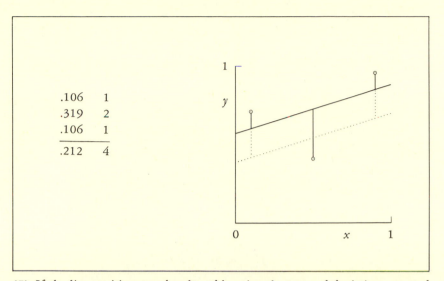

.106	1
.319	2
.106	1
.212	4

(5) If the line position now be altered keeping the two end deviations equated in magnitude (parallel displacement), it is then clear that $(d_0 + d_2) + 2d_1$ remains constant. From these considerations, we deduce the invariant relationship $d_0 + 2d_1 + d_2 = 4d$.

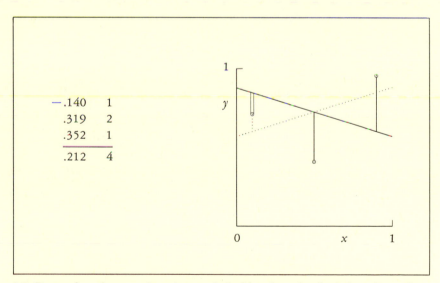

$-$.140	1
.319	2
.352	1
.212	4

(6) Remember that our invariant only holds when the deviations have signs appropriate to best fit. Should the law of signs be violated, however, the invariant can still be made to hold by treating the offending deviation's magnitude as if it were negative.

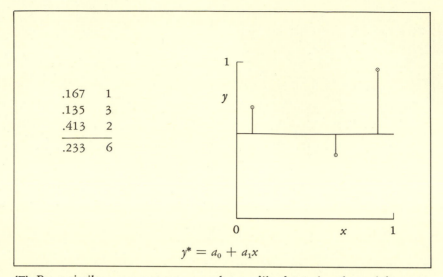

$$y^* = a_0 + a_1 x$$

(7) By a similar argument, we can also readily determine the weights associated with the fitting of a straight line to three points that are not equally spaced in the x-coordinate.

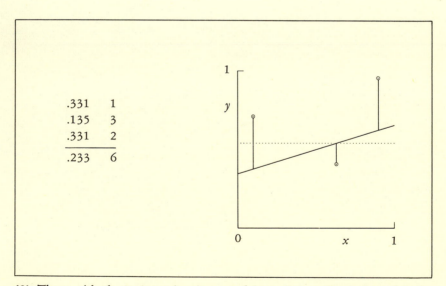

(8) Thus, with the $2:1$ spacing in x used here, it is quite obvious that the value of $d_0 + 2d_2$ must remain fixed when the middle deviation is held fixed, and again it is clear that the middle weight must be the sum of the other two.

	x	y^*	y	ε	
	.1		.7		
	.2		1.1		
	.3		1.4		
	.4		1.7		
	.5		2.2		
	.6		2.5		
	.7		2.9		
	.8		3.2		

(9) *Prologue to "An Iterative Procedure"*: With this brief introduction to the subject of weights, which will be elaborated upon in a later chapter, let us now consider the problem of best-fitting a straight line to the eight points listed on the printed form, above.

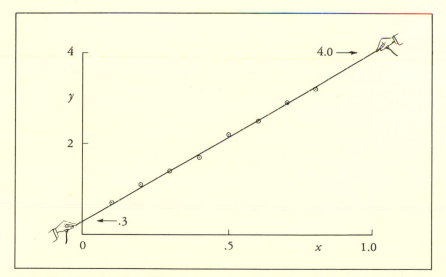

(10) We plot the data. The urge to use that piece of string once more is irresistible. And so, noting the intercepts at 0 and 1, we write $y^* = .3 + 3.7x$ as a first approximation to the required line of best fit.

120114

.3	3.7				
x	y^*	y	ε		
.1	.67	.7	$-.03$		
.2	1.04	1.1	$-.06$	?	
.3	1.41	1.4	.01		
.4	1.78	1.7	.08	?	
.5	2.15	2.2	$-.05$	?	
.6	2.52	2.5	.02		
.7	2.89	2.9	$-.01$		
.8	3.26	3.2	.06		

(11) Then we compute error data and thus obtain a closer look at the deviations. As we are fitting a linear form with two parameters, some three of the eight points will determine the final fit.

$$
\begin{array}{cc}
.06 & 1 \\
.08 & 3 \\
.05 & 2 \\
\hline
.067 & 6
\end{array}
$$

$$a_0 + .2a_1 = 1.1 - .067$$

$$a_0 + .5a_1 = 2.2 - .067$$

$$a_0 = .300$$

$$a_1 = 3.667$$

(12) We mark the three points that appear to be the significant ones and use the $1:3:2$ weights of Frame (8) to determine d, as our x-spacing is $2:1$. We set up and solve equations which impose deviations of magnitude d with proper signs at the two end points.

.300	3.667				
	x	*y**	*y*	*ε*	
	.1	.667	.7	−.033	
	.2	1.033	1.1	−.067	!
	.3	1.400	1.4	.000	
	.4	1.767	1.7	.067	!
	.5	2.134	2.2	−.066	!
	.6	2.500	2.5	.000	
	.7	2.867	2.9	−.033	
	.8	3.234	3.2	.034	

(13) Again we compute error data (we have quite a supply of these printed forms), and this time it is apparent that we have the desired line of best fit. We shall say that our error of approximation is $\varepsilon = .067$.

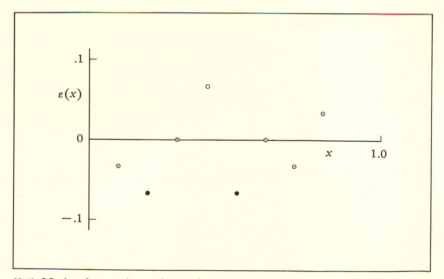

(14) Notice that we have obtained our best fit by imposing all but one of the equal greatest deviations in such a manner that the remaining equal greatest deviation comes up with the right magnitude.

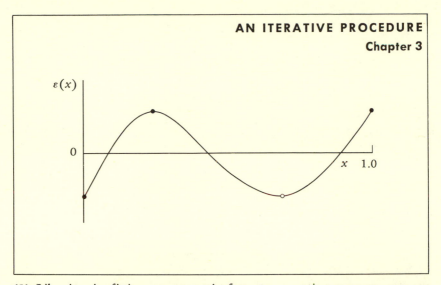

(1) Likewise, in fitting a parametric form to a continuous curve over an interval, we obtain our final fit by imposing all but one of the greatest equal deviations in such a manner that the remaining equal greatest deviation comes up with the right magnitude.

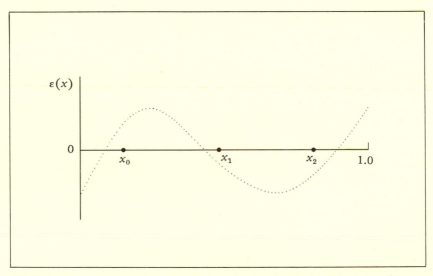

(2) But how do we get in position to take this final step? Imagine that we are fitting a parametric form containing three parameters to a segment of curve over $(0, 1)$. We might begin by guessing at the three root locations of the final, leveled, but (alas!) unknown, error curve.

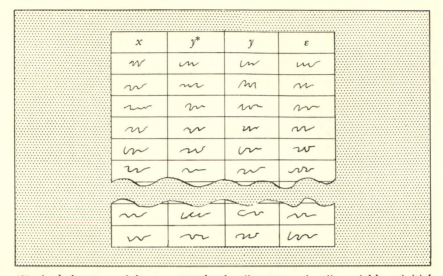

(3) And then we might set up and solve "root equations" to yield an initial approximation. Having such an initial approximation, we next compute error-curve plotting data.

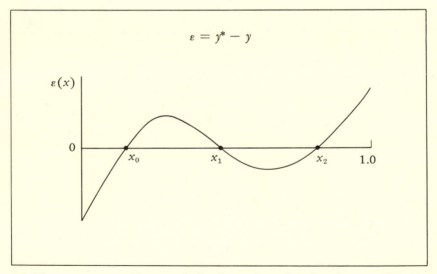

(4) The error-curve points so obtained are then plotted and a smooth curve is drawn through them. (Usually we plot about sixty or seventy points to obtain a single error curve!) Naturally the curve drawn has the roots imposed, if our numerical work has been done properly.

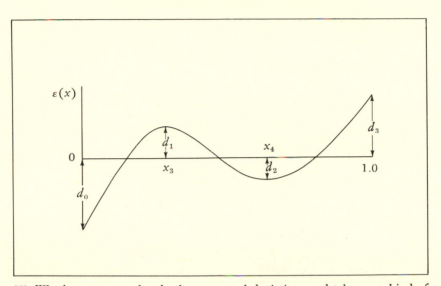

(5) We then measure the absolute extremal deviations and take some kind of a weighted average of the d_i to obtain an estimate of d. If we have an idea as to what kind of weights to use, fine! If not, we may try taking a simple average.

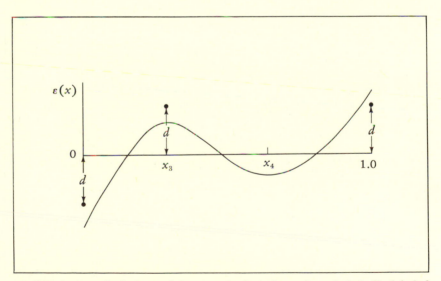

(6) We impose deviations of the magnitude estimated, optimistically labeled d, above, with proper sign at three of the observed extremal locations. We say, also optimistically, that we are imposing extremals and call the resulting equations "extremal equations."

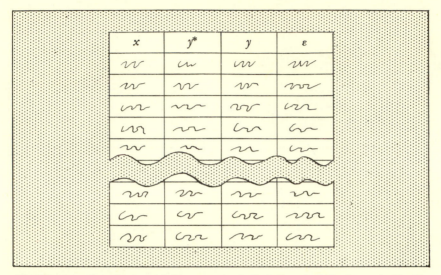

(7) We say that the remaining extremal deviation is left free. Again the equations are solved, and new error-curve data are computed.

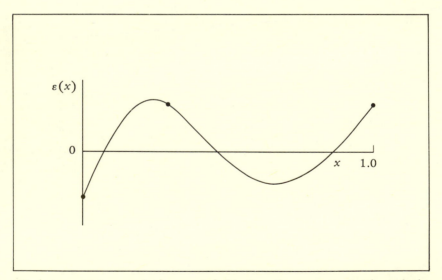

(8) A new error curve is drawn, and this time the curve passes through the points decided upon in Frame (6), if our numerical work has been done correctly.

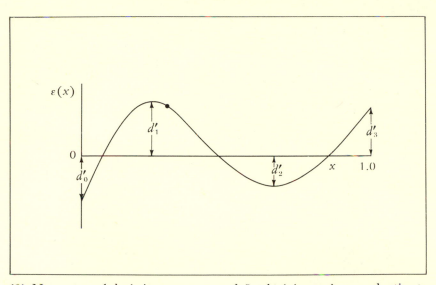

(9) New extremal deviations are measured. In obtaining an improved estimate of d, we use all extremal data available and improve upon our estimates of the weights, if possible.

	First Cycle		Second Cycle
	$d_0 \quad w_0$		$d_0' \quad w_0$
	$d_1 \quad w_1$		$d_1' \quad w_1$
	$d_2 \quad w_2$		$d_2' \quad w_2$
	$d_3 \quad w_3$		$d_3' \quad w_3$
	$d \quad w$		$d' \quad w$

(10) Roughly speaking, we try to find a set of positive weights that give nearly identical estimates of d for both sets of deviations. Some theory and advice on this subject are given in a later chapter on weights.

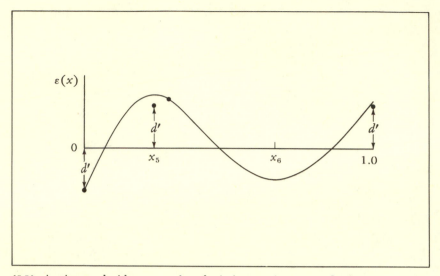

(11) Again we decide upon what deviations to impose and where.

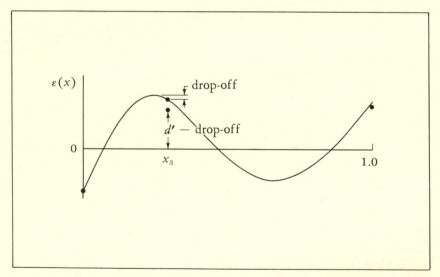

(12) Sometimes we save a little numerical labor by imposing reduced deviations at previously used locations. We may thus avoid setting up and solving equations entirely from scratch.

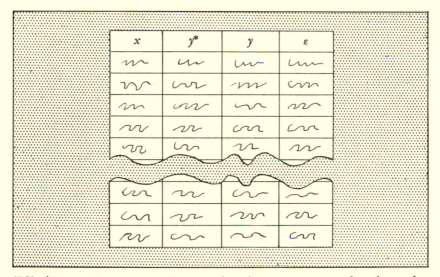

(13) At any rate, we set up our equations in one manner or the other, solve them, and again obtain plotting data. In accordance with the first alternative, the leveled error curve of Frame (1) is obtained.

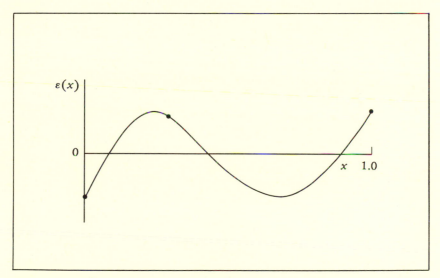

(14) In accordance with the second alternative, the nearly level error curve, above, is obtained. Generally, convergence is not quite as rapid as it might otherwise be if we try imposing deviations at all far from the true extremal locations.

Not This

$$\frac{1 + a_1 z_i}{2 + b_1 z_i + b_2 z_i^2 + b_3 z_i^3} = w(z_i)$$

◆　　　◆　　　◆　　　◆

$$\frac{1 + .4a_1}{2 + .4b_1 + .16b_2 + .064b_3} = .37552$$

$$\frac{1 + 1.8a_1}{2 + 1.8b_1 + 3.24b_2 + 5.832b_3} = .16155$$

(1) Virtually all the parametric forms that we shall care to consider involve only the solution of linear equations in the process of fitting. The only forms considered in this monograph leading to nonlinear equations are those on Sheets 43, 44, and 45.

But This

$$-a_1 + b_1 w(z_i) + b_2 z_i w(z_i) + b_3 z_i^2 w(z_i) = \frac{1 - 2w(z_i)}{z_i}$$

◆　　　◆　　　◆　　　◆

$$\dot{a}_1 + .37552 b_1 + .150208 b_2 + .0600832 b_3 = .62240000$$

$$\dot{a}_1 + .16155 b_1 + .290790 b_2 + .5234220 b_3 = .37605556$$

$$\dot{a}_1 + .03803 b_1 + .197756 b_2 + 1.0283312 b_3 = .17768077$$

$$\dot{a}_1 + .01122 b_1 + .109956 b_2 + 1.0775688 b_3 = .09975102$$

$$(\dot{a}_1 = -a_1)$$

(2) Of course, we have to write our type equation down properly or the resulting systems of equations to be solved may appear to be nonlinear; but this is a simple task for any case considered herein, save the three exceptions mentioned above.

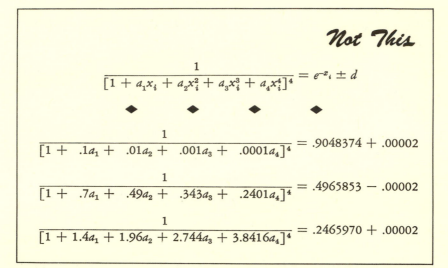

Not This

$$\frac{1}{[1 + a_1x_i + a_2x_i^2 + a_3x_i^3 + a_4x_i^4]^4} = e^{-x_i} \pm d$$

♦ ♦ ♦ ♦

$$\frac{1}{[1 + .1a_1 + .01a_2 + .001a_3 + .0001a_4]^4} = .9048374 + .00002$$

$$\frac{1}{[1 + .7a_1 + .49a_2 + .343a_3 + .2401a_4]^4} = .4965853 - .00002$$

$$\frac{1}{[1 + 1.4a_1 + 1.96a_2 + 2.744a_3 + 3.8416a_4]^4} = .2465970 + .00002$$

(3) This being the case, we are in business if we know a good method for solving systems of linear equations. We can think of no better method for everyday use than the procedure devised by Dr. Prescott D. Crout, of MIT.

But This

$$a_1 + a_2x_i + a_3x_i^2 + a_4x_i^3 = \frac{(e^{-x_i} \pm d)^{-\frac{1}{4}} - 1}{x_i}$$

♦ ♦ ♦ ♦

$$a_1 + .1a_2 + .01a_3 + .001a_4 = .2530946$$
$$a_1 + .7a_2 + .49a_3 + .343a_4 = .2732260$$
$$a_1 + 1.4a_2 + 1.96a_3 + 2.744a_4 = .2993134$$
$$a_1 + 3.5a_2 + 12.25a_3 + 42.875a_4 = .3997922$$

(4) Notice that in setting up an arbitrary type equation intelligently, we assign unit coefficients to the leftmost unknown. To avoid treating such a special case, however, we'll not be as clever as usual in setting up an example to work numerically.

Example

$$a_1 x_i + a_2 x_i^2 + a_3 x_i^3 + a_4 x_i^4 = \ln(1 + x_i)$$

◆   ◆   ◆   ◆

$$.2a_1 + .04a_2 + .008a_3 + .0016a_4 = .18232 \qquad .43192$$
$$.5a_1 + .25a_2 + .125a_3 + .0625a_4 = .40547 \qquad 1.34297$$
$$.8a_1 + .64a_2 + .512a_3 + .4096a_4 = .58779 \qquad 2.94939$$
$$1.0a_1 + 1.00a_2 + 1.000a_3 + 1.0000a_4 = .69315 \qquad 4.69315$$

(5) With this thought in mind, we now consider the above system of equations. As a first step, we sum the numbers in each row to obtain a "check column," which will be treated throughout as if it were an alternative right-hand column of numbers.

$$a_1 + .2\,a_2 + .04\,a_3 + .008\,a_4 = .91160 \qquad 2.15960$$
$$.5a_1 + .25a_2 + .125a_3 + .0625a_4 = .40547 \qquad 1.34297$$
$$.8a_1 + .64a_2 + .512a_3 + .4096a_4 = .58779 \qquad 2.94939$$
$$\qquad\qquad\qquad\qquad\qquad\qquad\qquad 315$$

$$a_1 + .2\,a_2 + .04\,a_3 + .008\,a_4 = .91160 \qquad 2.15960$$
$$.15a_2 + .105a_3 + .0585a_4 = -.05033 \qquad .26317$$
$$.48a_2 + .48\,a_3 + .4032a_4 = -.14149 \qquad 1.22171$$

$$a_1 + .2\,a_2 + .04\,a_3 + .008\,a_4 = .91160 \qquad 2.15960$$
$$a_2 + .7\,a_3 + .39\,a_4 = -.33553 \qquad 1.75447$$
$$.48a_2 + .48\,a_3 + .4032a_4 = -.14149 \qquad 1.22171$$
$$.8\,a_2 + .96\,a_3 + .992\,a_4 = -.21845 \qquad 2.53355$$

(6) We divide the first equation through by its leading coefficient. The new first equation is then multiplied by each other leading coefficient in turn, and each result is subtracted from the corresponding original equation to yield a new equation sans an a_1 term. The process is continued.

$$
\begin{aligned}
.2a_1 &+ .04a_2 + .008a_3 + .0016a_4 = & .18232 & \quad .43192 \\
.5a_1 &+ .25a_2 + .125a_3 + .0625a_4 = & .40547 & \quad 1.34297 \\
.8a_1 &+ .64a_2 + .512a_3 + .4096a_4 = & .58779 & \quad 2.94939 \\
1.0a_1 &+ 1.00a_2 + 1.000a_3 + 1.0000a_4 = & .69315 & \quad 4.69315 \\[6pt]
a_1 &+ .2\ a_2 + .04\ a_3 + .008\ a_4 = & .91160 & \quad 2.15960 \\
& a_2 + .7\ a_3 + .39\ a_4 = & -.33553 & \quad 1.75447 \\
& a_3 + 1.5\ a_4 = & .13586 & \quad 2.63586 \\
& a_4 = & -.05462 & \quad .94538
\end{aligned}
$$

$$
\begin{array}{cccc}
.99666 & -.46668 & .21779 & -.05462 \\
1.99666 & .53332 & 1.21779 & .94538
\end{array}
$$

(7) In this fashion, we reduce the original equations to triangular form. At every stage of the numerical work, the computed check entry must agree with the sum of coefficients plus right member on the same line. Finally, the unknowns and their check values are computed in obvious fashion.

.2	.04	.008	.0016	.18232	.43192
.5	.25	.125	.0625	.40547	1.34297
.8	.64	.512	.4096	.58779	2.94939
1.0	1.00	1.000	1.0000	.69315	4.69315
.2	.2	.04	.008	.91160	2.15960
.5	.15	.7	.39	−.33553	1.75447
.8	.48	.144	1.5	.13586	2.63586
1.0	.8	.4	.08	−.05462	.94538
.99666	−.46668	.21779	−.05462		
1.99666	.53332	1.21779	.94538		

(8) Dr. Crout's procedure consists in a convenient organization of all this numerical labor, as illustrated above. The first column of the system matrix is copied to yield the first column of the auxiliary matrix, and the calculation of the first row of the auxiliary matrix is obvious.

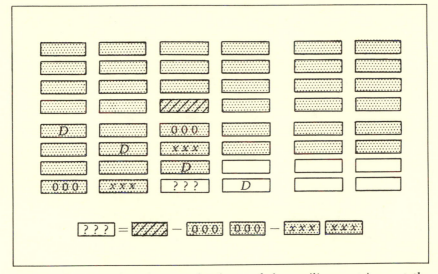

(9) We then complete the second column of the auxiliary matrix; next the second row, and so forth. The diagram above illustrates the general operation by which each diagonal entry or entry below a diagonal entry is computed.

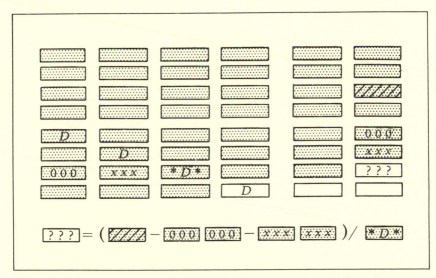

(10) And the diagram above illustrates the manner in which each entry to the right of a diagonal entry is computed. The two operations differ only in that we divide by the diagonal element on the line when it is to the left of the number being computed.

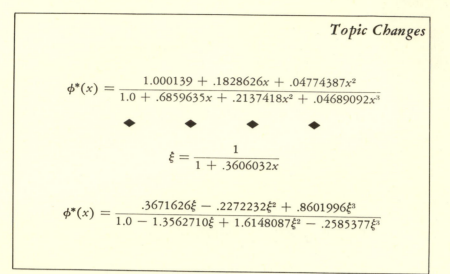

$$\phi^*(x) = \frac{1.000139 + .1828626x + .04774387x^2}{1.0 + .6859635x + .2137418x^2 + .04689092x^3}$$

◆   ◆   ◆   ◆

$$\xi = \frac{1}{1 + .3606032x}$$

$$\phi^*(x) = \frac{.3671626\xi - .2272232\xi^2 + .8601996\xi^3}{1.0 - 1.3562710\xi + 1.6148087\xi^2 - .2585377\xi^3}$$

(11) In several instances we have rewritten our original approximations in "desensitized" form to facilitate their evaluation on machines (such as the ENIAC) which compute to a fixed number of decimal places. Thus, the approximation of Sheet 6 is a restatement of the top-line expression above.

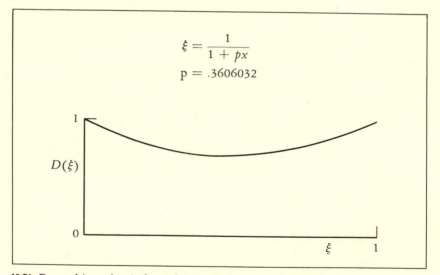

$$\xi = \frac{1}{1 + px}$$

$$p = .3606032$$

$D(\xi)$

(12) By making the indicated transformation, the interval $0 \le x < \infty$ is reduced to $0 \le \xi \le 1$, and the variations of the transformed numerator $N(\xi)$ and denominator $D(\xi)$ become bounded rather than unbounded.

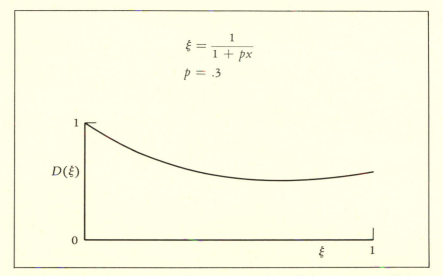

$$\xi = \frac{1}{1 + px}$$

$$p = .3$$

$D(\xi)$

(13) In quite general circumstances, the transformed denominator will have minimum variation when p is chosen to make $D(1) = D(0)$. In the example considered here, a smaller value of p gives a lesser $D(1)$, and the variation of $D(\xi)$ is seen to be increased.

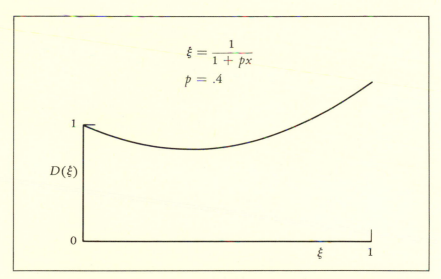

$$\xi = \frac{1}{1 + px}$$

$$p = .4$$

$D(\xi)$

(14) Conversely, a larger value of p gives a greater $D(1)$, and again the variation of the denominator is seen to be increased. In the situation pictured here, is it not quite reasonable to expect that the minimum $D(\xi)$ value will increase more sluggishly than the $D(1)$ value?

$$\Phi^*(x) = 1 - \left[\frac{a_1}{(1 + px)} + \frac{a_2}{(1 + px)^2} + \frac{a_3}{(1 + px)^3}\right]\Phi'(x)$$

$$\Phi^*(x) = 1 - \left[\frac{c_0 + c_1 x + c_2 x^2}{(1 + px)^3}\right]e^{-x^2}$$

(15) Let us now consider the fitting of Sheet 43, in which we do run into nonlinear equations. We begin by rewriting the parametric form considered in terms of new unknowns c_0, c_1, and c_2. This allows us to write the equations to be solved in as linear a form as possible.

$$y(x) = e^{x^2}\left[1 - \frac{2}{\sqrt{\pi}} \int_0^x e^{-t^2}\, dt\right]$$

$$c_0 + c_1 x_0 + c_2 x_0^2 = (1 + px_0)^3 y(x_0)$$
$$c_0 + c_1 x_1 + c_2 x_1^2 = (1 + px_1)^3 y(x_1)$$
$$c_0 + c_1 x_2 + c_2 x_2^2 = (1 + px_2)^3 y(x_2)$$
$$c_0 + c_1 x_3 + c_2 x_3^2 = (1 + px_3)^3 y(x_3)$$

(16) A set of "root equations" would then be written as above. Our problem is that of solving for unknowns c_0, c_1, c_2, and p. We begin by noting that p must be assigned a value p_0 that will make the four equations above consistent in the unknowns c_0, c_1, and c_2.

$$f(p) = \begin{vmatrix} 1 & x_0 & x_0^2 & (1 + px_0)^3 y(x_0) \\ 1 & x_1 & x_1^2 & (1 + px_1)^3 y(x_1) \\ 1 & x_2 & x_2^2 & (1 + px_2)^3 y(x_2) \\ 1 & x_3 & x_3^2 & (1 + px_3)^3 y(x_3) \end{vmatrix}$$

(17) These equations will be consistent if p_0 is chosen to make the above determinant equal to zero. To determine this number, we expand the determinant by elements of the rightmost column. We readily see that $f(p)$ is a cubic polynomial in p.

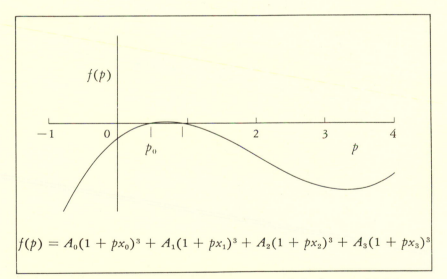

$$f(p) = A_0(1 + px_0)^3 + A_1(1 + px_1)^3 + A_2(1 + px_2)^3 + A_3(1 + px_3)^3$$

(18) But not being overly fond of algebra, we refrain from multiplying out the third-power terms. We leave the expression as above, and then use a standard iterative procedure, such as Newton's rule, to determine p_0.

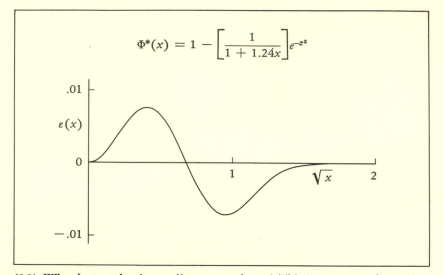

$$\Phi^*(x) = 1 - \left[\frac{1}{1 + 1.24x}\right]e^{-x^2}$$

(19) Why do we take the smallest root to be p_0? This is a very good question, and we don't have the best answer in the world. We began by trying to fit the $n = 1$ case of the parametric form under consideration. A nearly level error curve was obtained for $p_0 = 1.24$.

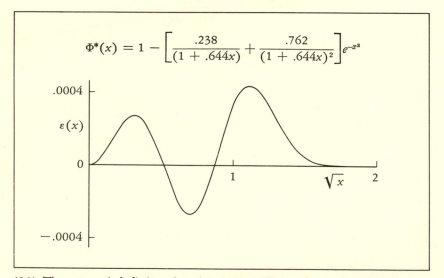

$$\Phi^*(x) = 1 - \left[\frac{.238}{(1 + .644x)} + \frac{.762}{(1 + .644x)^2}\right]e^{-x^2}$$

(20) Then we tried fitting the $n = 2$ case. We set up equations imposing reasonable error-curve roots, and this led us to solve a quadratic equation in p, yielding roots .644 and 2.18. The smaller root gave the better result. We then felt that the p_0 we wanted should decrease with increasing n.

$$c_0 + c_1 x_0 + c_2 x_0^2 = (1 + p_0 x_0)^3 y(x_0)$$
$$c_0 + c_1 x_1 + c_2 x_1^2 = (1 + p_0 x_1)^3 y(x_1)$$
$$c_0 + c_1 x_3 + c_2 x_3^2 = (1 + p_0 x_3)^3 y(x_3)$$

(21) And so we choose to call the smallest root p_0. Returning to the equations of Frame (16), we replace p by p_0 and then discard any one of the four equations. The remaining three linear equations are then solved for the unknowns c_0, c_1, and c_2.

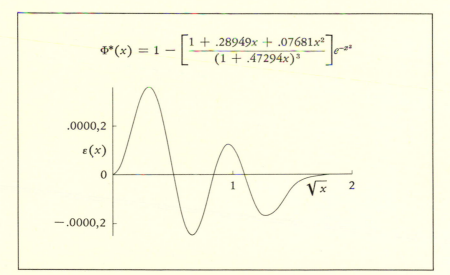

$$\Phi^*(x) = 1 - \left[\frac{1 + .28949x + .07681x^2}{(1 + .47294x)^3} \right] e^{-x^2}$$

(22) The graph above and that of Frame (18) pertain to an actual cycle in the fitting of Sheet 43. Roots were imposed at $x = 0$, .25, .7, and 1.2. The above error curve was obtained and then leveled in later cycles.

REFERENCES

1. CROUT, PRESCOTT D., "A Short Method for Evaluating Determinants and Solving Systems of Linear Equations with Real or Complex Coefficients," Marchant Methods, MM-182 Mathematics, September, 1941.
2. MILNE, WILLIAM E., *Numerical Calculus*, Princeton University Press, Princeton, N.J., 1949.

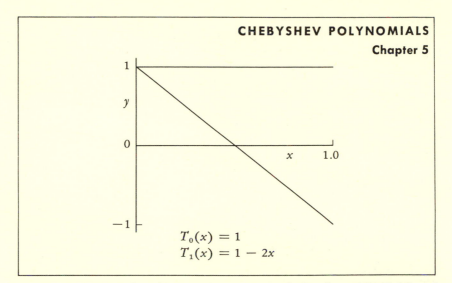

$$T_0(x) = 1$$
$$T_1(x) = 1 - 2x$$

(1) We shall now describe an infinite sequence of polynomials $T_n(x)$, of exact degree n in x, called the Chebyshev polynomials of the first kind. Their theory throws much interesting light on the subject of polynomial approximation.

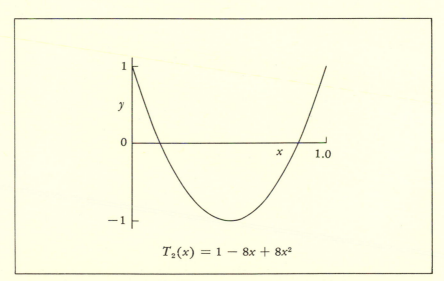

$$T_2(x) = 1 - 8x + 8x^2$$

(2) To begin with, let us say that we may think of each polynomial $T_n(x)$ as having been constructed to start at the point $(0, 1)$ and to oscillate back and forth between ± 1 as many times as possible for a polynomial of its degree in the interval $(0, 1)$.

47

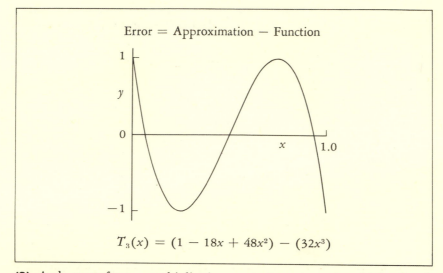

Error = Approximation − Function

$$T_3(x) = (1 - 18x + 48x^2) - (32x^3)$$

(3) And, apart from a multiplicative constant, we may also think of each $T_n(x)$ as being the equation of the error curve that results from best-fitting a polynomial of degree $n - 1$ to x^n over $(0, 1)$.

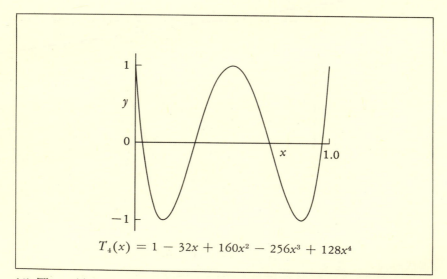

$$T_4(x) = 1 - 32x + 160x^2 - 256x^3 + 128x^4$$

(4) Thus, $T_4(x) = (1 - 32x + 160x^2 - 256x^3) - (-128x^4)$ in accordance with the formula "Error = Approximation − Function." As $a_0 + a_1x + a_2x^2 + a_3x^3$ is a *four*-parameter form, and the above curve has *five* equal greatest deviations of alternating sign, it is apparent that $T_4(x)$ has the properties of a "normal" error curve.

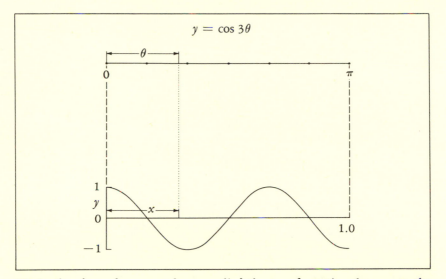

$$y = \cos 3\theta$$

(5) In the above frame, we have applied the transformation $\theta = \pi x$ to the function $y = \cos 3\theta$. We think of this one-to-one correspondence as defined by vertical projection to the lower axis. The root and extremal locations of $y = \cos 3\theta$ have been indicated on the θ-axis.

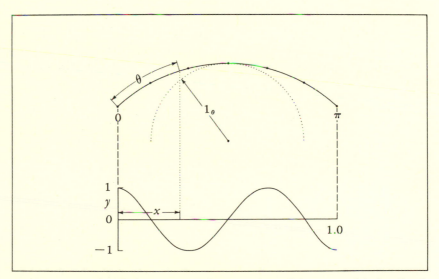

(6) Let us now bend the segment of θ-axis pictured into a semicircle of unit radius: the unit referred to here is that of the θ-axis scale. Naturally this changes the relationship between x and θ, which we continue to define by vertical projection.

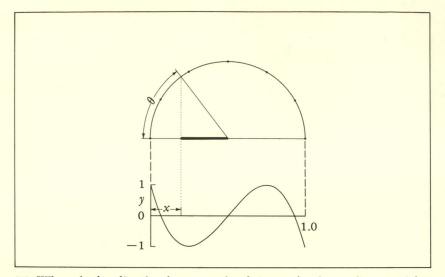

(7) When the bending has been completed, it may be shown that $\cos 3\theta$ has been warped into $T_3(x)$. But first, what is the present relationship between x and θ? As the radius of the semicircle is .5 in units of x, the accentuated line segment has a length of .5 $\cos \theta$ in units of x.

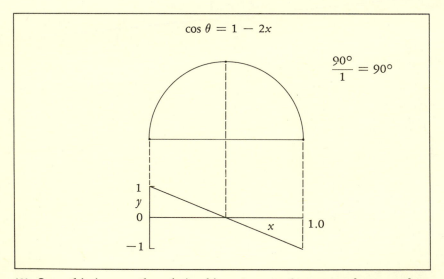

(8) Once this is seen, the relationship $x + .5 \cos \theta = .5$ stands out, and we next learn that our transformation can be written $\cos \theta = 1 - 2x$, which in turn says that $\cos \theta$ is transformed into $T_1(x)$.

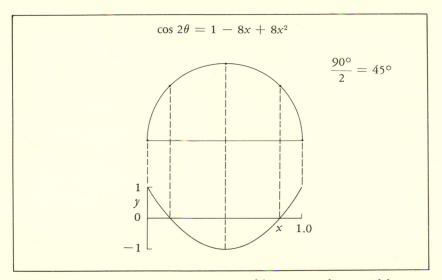

$$\cos 2\theta = 1 - 8x + 8x^2$$

$$\frac{90°}{2} = 45°$$

(9) Then, as $\cos 2\theta = 2\cos^2 \theta - 1$, we readily compute that $\cos 2\theta$ is transformed into $2(1 - 2x)^2 - 1 = 1 - 8x + 8x^2 = T_2(x)$. *Now, if we can prove that* $\cos n\theta$ *transforms under* $\cos \theta = 1 - 2x$ *into a polynomial of degree n in x, then that polynomial must be* $T_n(x)$.

$$\cos (n + 1)\theta = \cos n\theta \cos \theta - \sin n\theta \sin \theta$$

and

$$\cos (n - 1)\theta = \cos n\theta \cos \theta + \sin n\theta \sin \theta$$

when added yield

$$\cos (n + 1)\theta = 2 \cos \theta \cos n\theta - \cos (n - 1)\theta$$

which transforms to

$$T_{n+1}(x) = 2(1 - 2x)T_n(x) - T_{n-1}(x)$$

(10) *This is because* $\cos n\theta$ *oscillates between* ± 1 *the appropriate number of times, and this property is preserved under transformation. Furthermore, the transformed polynomial takes on the required value at* $x = 0$.

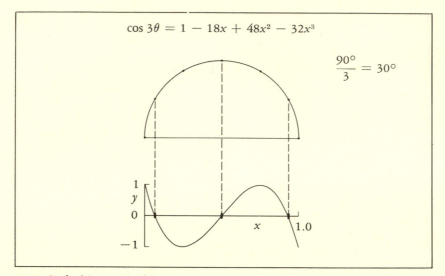

$$\cos 3\theta = 1 - 18x + 48x^2 - 32x^3$$

$$\frac{90°}{3} = 30°$$

(11) And this we readily show by induction. Thus, the identity $\cos 3\theta = 2 \cos \theta \cos 2\theta - \cos \theta$, coupled with the fact that $\cos \theta = 1 - 2x$ and $\cos 2\theta = 1 - 8x + 8x^2$ are polynomials of required degree, shows that $\cos 3\theta$ transforms to a polynomial of third degree.

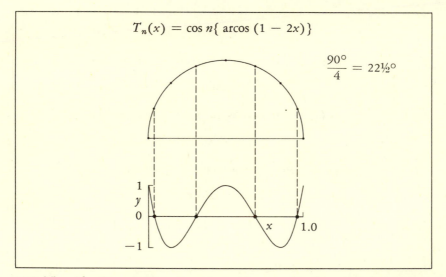

$$T_n(x) = \cos n\{ \arccos (1 - 2x)\}$$

$$\frac{90°}{4} = 22\frac{1}{2}°$$

(12) The identity $\cos 4\theta = 2 \cos \theta \cos 3\theta - \cos 2\theta$ then transforms to a polynomial of fourth degree, and the induction may be carried out without difficulty. Finally, let us remark that, as $\theta = \arccos (1 - 2x)$, we may write the above simple formula for $T_n(x)$ valid over $0 \leq x \leq 1$.

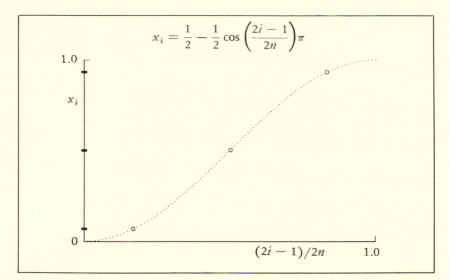

$$x_i = \frac{1}{2} - \frac{1}{2}\cos\left(\frac{2i-1}{2n}\right)\pi$$

(13) Now, in the preceding sequence of diagrams, we emphasize, by vertical projection from the semicircle, the manner in which the roots (and extremal locations) of $\cos n\theta$ are transformed into the roots (and extremal locations) of $T_n(x)$.

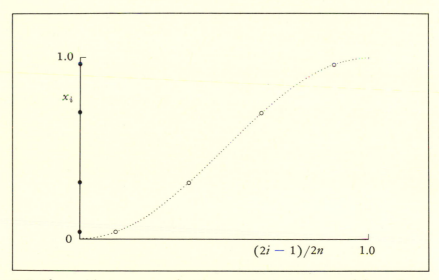

(14) If we wish, we may number the roots $i = 1, 2, \cdots, n$ and draw a "root-location curve." That is, the roots of $T_n(x)$ must satisfy the equation $n \arccos (1 - 2x_i) = (2i - 1)\pi/2$ rewritten above. The idea of looking for such curves in nonpolynomial approximation is a rather interesting one!

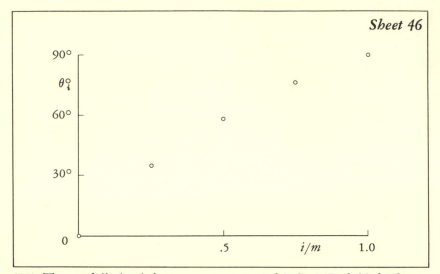

(15) The carefully leveled error curves presented in Part II of this book contain some, but all too little, data to be analyzed on the subject of root and extremal locations. Consider Sheet 46, for example. We number the five roots $i = 0, 1, 2, 3, 4$ and plot root values $\theta_i^{\circ} = \arcsin k_i$ against $i/4$ to yield the plot above.

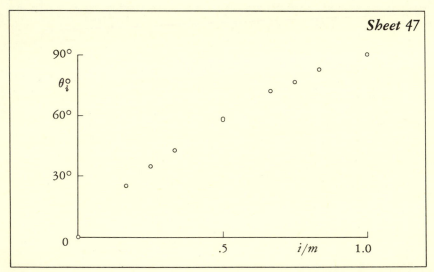

(16) We next consider Sheet 47, which contains the second approximation in the sequence. We number the roots $i = 0, 1, \cdots, 6$ and plot θ_i° against $i/6$ on the same sheet of graph paper. It now appears that our roots lie roughly on a curve much as did the roots of $T_n(x)$.

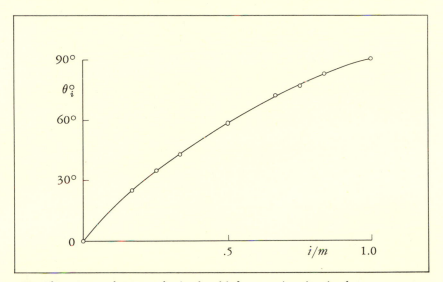

(17) If we were about to obtain the third approximation in the sequence, we should now surely draw an approximate root-location curve and read off θ_i° values from this curve for $i/m = 0(\frac{1}{8})1$. These θ_i° values would be used to initiate the iterative procedure of Chapter 3.

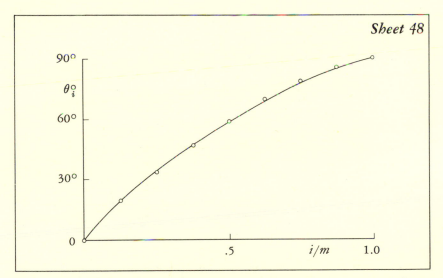

(18) And we shouldn't be badly disappointed! In conclusion, it is very interesting to note that the error curves of Sheets 49, 50, and 51 fall *almost exactly* on top of the corresponding error curves for Sheets 46, 47, and 48.

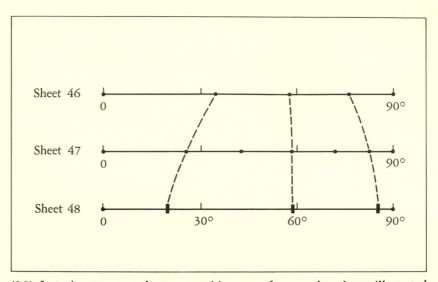

(19) In trying to extrapolate root positions we often use the scheme illustrated here. Imagine that Sheet 48 root locations must be estimated. Two roots are being added at each step; the position of the middle root appears to be quite stable. So we extrapolate the middle-root position.

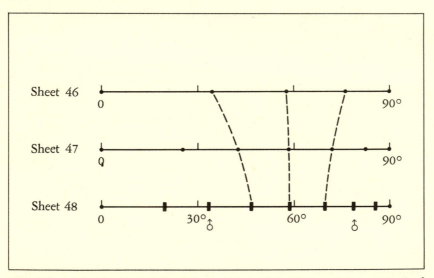

(20) We might then extrapolate the position of the first and last roots. And then we might extrapolate the positions of the roots to either side of the middle root. Finally we might fudge in (○→), the two other missing root locations.

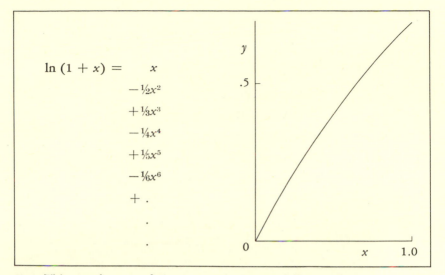

$$\ln(1+x) = \quad x$$
$$-\tfrac{1}{2}x^2$$
$$+\tfrac{1}{3}x^3$$
$$-\tfrac{1}{4}x^4$$
$$+\tfrac{1}{5}x^5$$
$$-\tfrac{1}{6}x^6$$
$$+\ .$$

(21) This may be a good time to emphasize the difference between a "point expansion" and an "interval expansion." The usual power-series development may be called a "point expansion," in that the successive terms of the series match the ordinate and successive derivatives of the function at some one specified point.

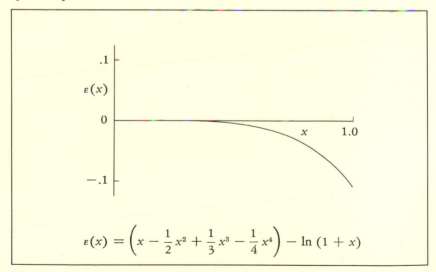

$$\varepsilon(x) = \left(x - \frac{1}{2}x^2 + \frac{1}{3}x^3 - \frac{1}{4}x^4\right) - \ln(1+x)$$

(22) When such an expansion is curtailed—that is, when all terms beyond a certain point are discarded—an approximation is obtained which has all its disposable roots located at one point and has only accidental roots elsewhere.

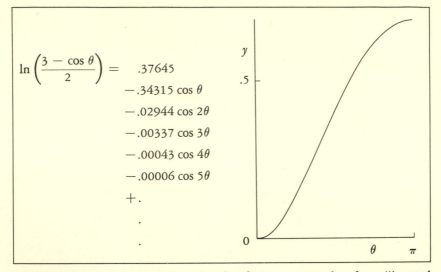

$$\ln\left(\frac{3 - \cos\theta}{2}\right) = \ .37645$$
$$-.34315 \cos\theta$$
$$-.02944 \cos 2\theta$$
$$-.00337 \cos 3\theta$$
$$-.00043 \cos 4\theta$$
$$-.00006 \cos 5\theta$$
$$+\,.$$
$$\cdot$$
$$\cdot$$

(23) A Fourier series, on the other hand, is an example of an "interval expansion," in that the determination of each parameter a_n is made to depend on all values of the function approximated within the interval of approximation.

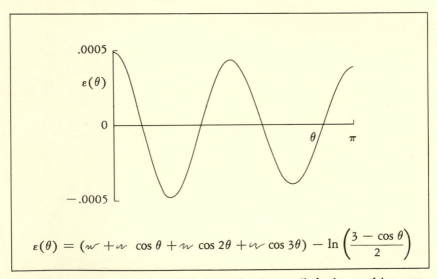

$$\varepsilon(\theta) = (\omega + \omega \cos\theta + \omega \cos 2\theta + \omega \cos 3\theta) - \ln\left(\frac{3 - \cos\theta}{2}\right)$$

(24) If a rapidly converging Fourier series is curtailed, the resulting error curve of approximation will have its roots quite evenly spaced. For, as in the present instance, the error of approximation is nearly given by the first cosine term neglected, except for sign which is reversed.

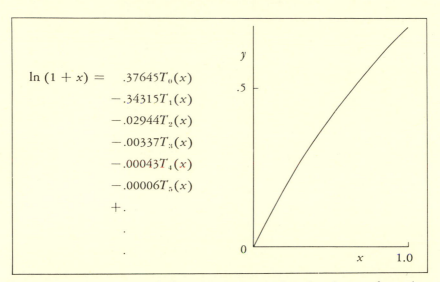

$$\ln(1 + x) = \quad .37645T_0(x)$$
$$- .34315T_1(x)$$
$$- .02944T_2(x)$$
$$- .00337T_3(x)$$
$$- .00043T_4(x)$$
$$- .00006T_5(x)$$
$$+ .$$

(25) From what has gone before, we now appreciate that the transformation $\cos \theta = 1 - 2x$ will convert a Fourier cosine series into an expansion in terms of Chebyshev polynomials, valid over the interval $(0, 1)$.

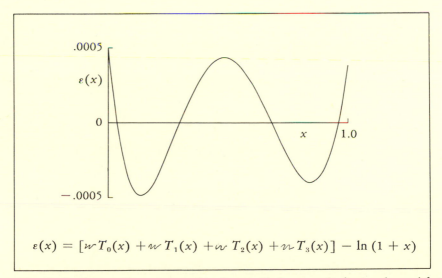

$$\varepsilon(x) = [\mathit{w}\, T_0(x) + \mathit{w}\, T_1(x) + \mathit{w}\, T_2(x) + \mathit{w}\, T_3(x)] - \ln(1 + x)$$

(26) And when such an expansion is curtailed, a nearly best polynomial approximation is obtained. The corresponding polynomial of best fit must therefore have an error curve with root and extremal locations much like those of the appropriate Chebyshev polynomial.

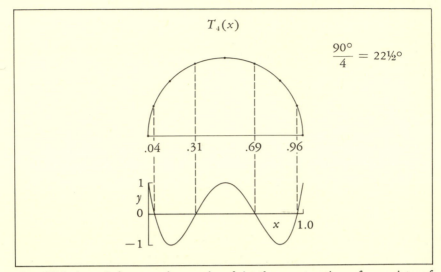

(27) This knowledge may be employed in the construction of a variety of work sheets for use in the fitting of polynomials, and of course it may be used to initiate the iterative procedure of Chapter 3—work sheet or no work sheet.

$$a_0 + a_1 x + a_2 x^2 + a_3 x^3 = y(x)$$

◆   ◆   ◆   ◆

$$a_0 + .04a_1 + .0016a_2 + .000064a_3 = y(.04)$$
$$a_0 + .31a_1 + .0961a_2 + .029791a_3 = y(.31)$$
$$a_0 + .69a_1 + .4761a_2 + .328509a_3 = y(.69)$$
$$a_0 + .96a_1 + .9216a_2 + .884736a_3 = y(.96)$$

(28) In a typical instance, we determine the root locations of the Chebyshev polynomial of appropriate order, set up root equations, and solve for the a_i in terms of the $y(x_i)$. In solving a specific problem, the $y(x_i)$ would be definite numbers; in constructing a work sheet, the $y(x_i)$ would remain as written above.

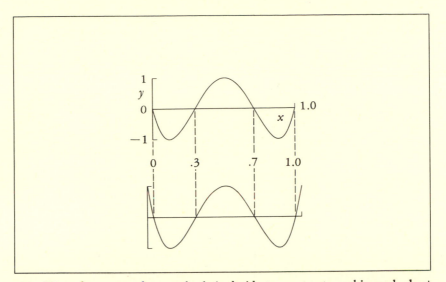

(29) Now, for a second example, let's decide to construct a cubic work sheet in which roots are to be imposed at each end of the interval $(0, 1)$. We chop the two ends from $T_4(x)$ and consider the reduced interval to be $(0, 1)$.

$$a_0 + a_1 x + a_2 x^2 + a_3 x^3 = y(x)$$

$$a_0 = y(0)$$
$$a_0 + .3a_1 + .09a_2 + .027a_3 = y(.3)$$
$$a_0 + .7a_1 + .49a_2 + .343a_3 = y(.7)$$
$$a_0 + a_1 + a_2 + a_3 = y(1.0)$$

(30) To an adequate degree of approximation, it suffices to record one-figure values of root and extremal locations. The general root equations are inverted, yielding $a_0 = y(0)$, $21a_1 = -121y(0) + 175y(.3) - 75y(.7) + 21y(1.0)$, etc.

Polynomial Work Sheet

$$y^*(x) = a_0 + a_1x + a_2x^2 + a_3x^3$$

		a_0	a_1	a_2	a_3
$y(0)$		1	-121	200	-100
$y(.3)$			175	-425	250
$y(.7)$			-75	325	-250
$y(1.0)$			21	-100	100
		1	21	21	21

		$x = .1$	$x = .5$	$x = .9$
a_0		1.0	1.0	1.0
a_1		.1	.5	.9
a_2		.01	.25	.81
a_3		.001	.125	.729
	y^*			
	y			
	ε			

(31) And here is the finished product, ready for use. The top portion of the work sheet is used to obtain the a_i, while the lower portion is used to test the resulting approximation at locations where the error is expected to be of extremal magnitude.

Polynomial Work Sheet

$$y^*(x) = a_0 + a_1 x + a_2 x^2 + a_3 x^3 + a_4 x^4$$

	a_0	a_1	a_2	a_3	a_4
$y(0)$	1	-333	972	-1125	225
$y(.2)$		500	-2125	2875	-625
$y(.5)$		-256	1856	-3200	800
$y(.8)$		125	-1000	2125	-625
$y(1.0)$		-36	297	-675	225
	1	36	36	36	18

	$x = .07$	$x = .34$	$x = .66$	$x = .93$
a_0	1.0	1.0	1.0	1.0
a_1	.07	.34	.66	.93
a_2	.0049	.1156	.4356	.8649
a_3	.0003,43	.0393,04	.2874,96	.8043,57
a_4	.0000,2401	.0133,6336	.1897,4736	.7480,5201
y^*				
y				
ε				

(32) The two sample work sheets given here are typical of dozens that we constructed for our own use in the fitting of polynomial and other types of parametric forms. Work sheets were constructed for use in imposing extremal deviations as well as for the imposing of roots.

REFERENCES

1. LANCZOS, CORNELIUS, "Trigonometric Interpolation of Empirical and Analytical Functions," *J. Math. Physics*, Vol. 17, No. 3, September, 1938, pp. 123–199.
2. National Bureau of Standards, *Tables of Chebyshev Polynomials $S_n(x)$ and $C_n(x)$*, U.S. Government Printing Office, Washington, D.C., 1952.

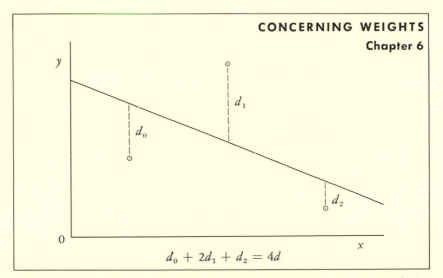

$$d_0 + 2d_1 + d_2 = 4d$$

(1) In a previous chapter, we employed simple geometrical reasoning to derive the invariant (shown above) associated with the fitting of a straight line to three points equally spaced in the x-coordinate.

$$a_0 + \qquad\quad x_0 a_1 = y_0 + d_0$$
$$a_0 + (x_0 + \ h)a_1 = y_1 - d_1$$
$$a_0 + (x_0 + 2h)a_1 = y_2 + d_2$$

◆   ◆   ◆   ◆

$$ha_1 = -y_0 + y_1 - d_0 - d_1$$
$$ha_1 = -y_1 + y_2 + d_1 + d_2$$

◆   ◆   ◆   ◆

$$d_0 + 2d_1 + d_2 = -y_0 + 2y_1 - y_2$$

(2) Let us now remark that this invariant can be derived algebraically in the manner illustrated here. The three deviations satisfy the top three equations. Elimination of a_0 and a_1 then yields the required invariant.

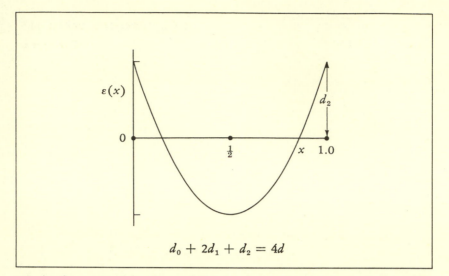

$$d_0 + 2d_1 + d_2 = 4d$$

(3) As the extremal locations of $T_2(x)$ are equally spaced, we shall say that weights $1:2:1$ are the Chebyshev weights associated with the fitting of $y^*(x) = a_0 + a_1x$.

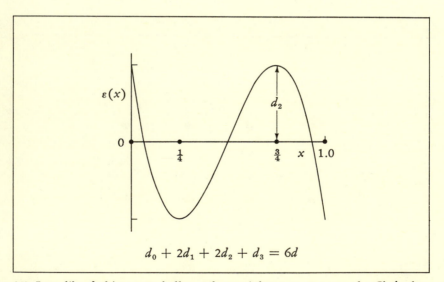

$$d_0 + 2d_1 + 2d_2 + d_3 = 6d$$

(4) In a like fashion, we shall say that weights $1:2:2:1$ are the Chebyshev weights associated with the fitting of $y^*(x) = a_0 + a_1x + a_2x^2$.

?

$$a_0 \qquad\qquad\qquad = y_0 + d_0$$
$$a_0 + \tfrac{1}{4} a_1 + \tfrac{1}{16} a_2 = y_1 - d_1$$
$$a_0 + \tfrac{3}{4} a_1 + \tfrac{9}{16} a_2 = y_2 + d_2$$
$$a_0 + \quad a_1 + \quad a_2 = y_3 - d_3$$

(5) The pertinent equations appear above. We suggest that the reader verify the equations and then eliminate the a_i to obtain the required invariant.

Chebyshev Weights

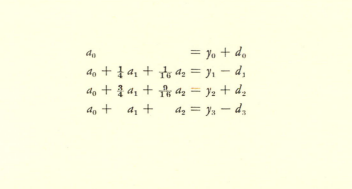

```
        1   1
     1   2   1
  1   2   2   1
 1   2   2   2   1
1   2   2   2   2   1
```

(6) In the actual process of fitting a polynomial, the extremal locations will shift a bit from cycle to cycle and probably will not end up where the theory indicates; but these are still the best a priori weights that we know of to use.

$$y^*(x_0:a, b, c) = y(x_0) + d$$
$$y^*(x_1:a, b, c) = y(x_1) - d$$
$$y^*(x_2:a, b, c) = y(x_2) + d$$
$$y^*(x_3:a, b, c) = y(x_3) - d$$

(7) The concept of weights is still a useful one in more general circumstances. Suppose that a best-fit situation is described by the equations above. The extremal locations are then at x_0, x_1, x_2, and x_3.

$$y^*(x_0:a + da, b + db, c + dc) = y(x_0) + d_0$$
$$y^*(x_1:a + da, b + db, c + dc) = y(x_1) - d_1$$
$$y^*(x_2:a + da, b + db, c + dc) = y(x_2) + d_2$$
$$y^*(x_3:a + da, b + db, c + dc) = y(x_3) - d_3$$

(8) A nearly best fit would then be described by these equations. Differential variations in x_i need not be taken into consideration, as the error curve is either flat at x_i or the extremal location is fixed.

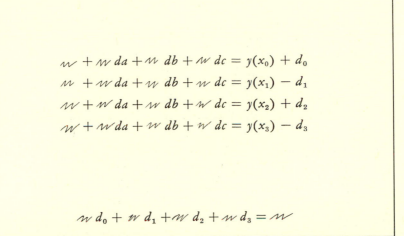

$$\sim + \sim da + \sim db + \sim dc = y(x_0) + d_0$$
$$\sim + \sim da + \sim db + \sim dc = y(x_1) - d_1$$
$$\sim + \sim da + \sim db + \sim dc = y(x_2) + d_2$$
$$\sim + \sim da + \sim db + \sim dc = y(x_3) - d_3$$

$$\sim d_0 + \sim d_1 + \sim d_2 + \sim d_3 = \sim$$

(9) Expanding each equation and neglecting higher-order terms, we get a system of consistent equations from which the differentials can be eliminated, leaving the indicated approximate invariant. *Note that the weights are not a function of the* $y(x_i)$.

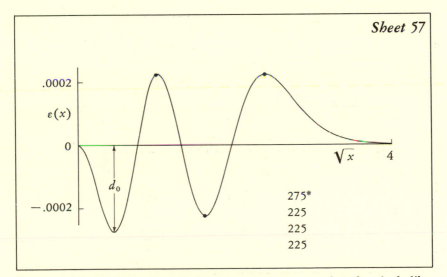

Sheet 57

$.0002$

$\varepsilon(x)$

0

$\sqrt{x}$ 4

d_0

275^*
225
225
225

$-.0002$

(10) Let's determine a few of these invariants to see what they look like. Taking Sheet 57, we leave d_0 free and squeeze the remaining extremal deviations down to .000225 (a quite arbitrary level), with the result shown above.

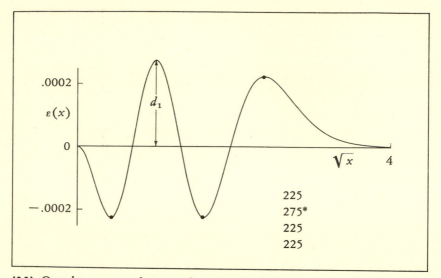

(11) One by one we leave each extremal deviation free and depress the remaining three peaks to the same value of .000225.

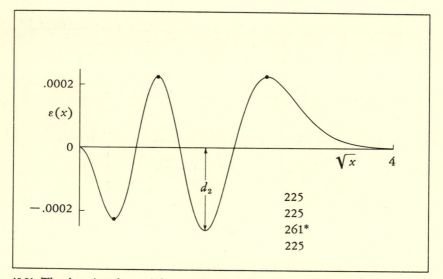

(12) The heavier the weight associated with a given deviation, the less the free peak stands above the others, and conversely.

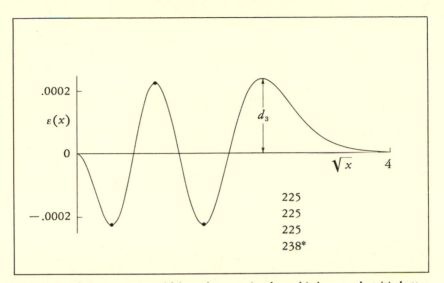

225
225
225
238*

(13) One of these cases could have been omitted, as d is known, but it's better to perturb them all.

Calculation of Weights

$$275w_0 + 225(1 - w_0) = d$$

$50w_0 = d - 225$	50	20	2
$50w_1 = d - 225$	50	20	2
$36w_2 = d - 225$	36	28	3
$13w_3 = d - 225$	13	77	8
		145	15

$$2d_0 + 2d_1 + 3d_2 + 8d_3 = 15d$$

(14) The resulting data are then used to solve for the weights appropriate to each deviation. (Save for the decimal point, 28 is the reciprocal of 36.)

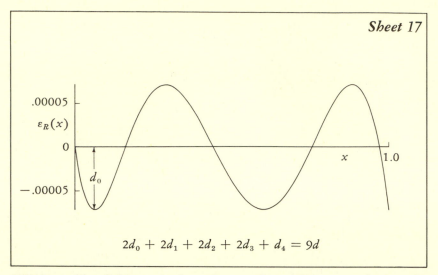

Sheet 17

$$2d_0 + 2d_1 + 2d_2 + 2d_3 + d_4 = 9d$$

(15) In a like fashion, we determine rough weights appropriate to the approximation of Sheet 17, where $[1 + a_1x + a_2x^2 + a_3x^3 + a_4x^4]^2$ is fitted to 10^x in the sense of relative error. How Chebyshev-like the weights are!

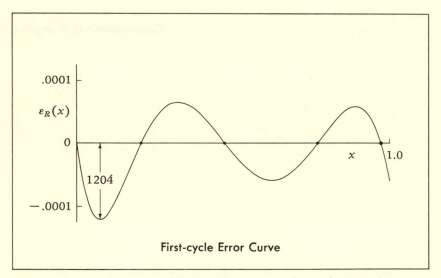

First-cycle Error Curve

(16) In the actual process of fitting Sheet 17, we began by imposing roots, as illustrated above. Extremal deviations were then measured, and, as we thought that the leftmost peak should be weighted quite heavily, we considerably overestimated d.

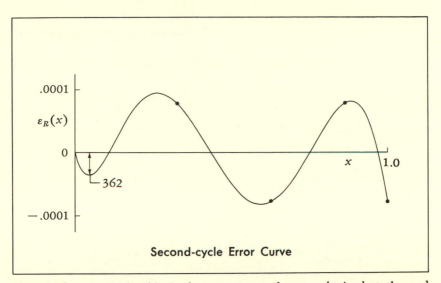

Second-cycle Error Curve

(17) And, as a result, this is the error curve that we obtained at the end of the second cycle. In practice, we usually only need to have an idea of the weight of the free extremal in order to do a pretty efficient job of fitting.

First Cycle		Second Cycle	
1204		362	
659 ⎤		952 ⎤	
584 ⎟		816 ⎟	
585 ⎟		813 ⎟	
590 ⎦		783 ⎦	
1204	.22 $\doteq \frac{2}{9}$	362	.22
604	.78	841	.78
736		736	

(18) We often estimate this number by "lumping," as illustrated here. The bracketed deviations are replaced by the simple average and the weights are considered to be lumped. Then we solve for a value of w_0 (in this case) that gives like estimates of d for the two sets of data.

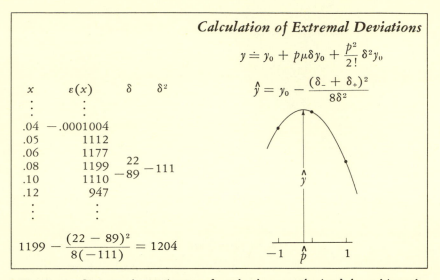

Calculation of Extremal Deviations

$$y \doteq y_0 + p\mu\delta y_0 + \frac{p^2}{2!}\delta^2 y_0$$

$$\hat{y} = y_0 - \frac{(\delta_- + \delta_+)^2}{8\delta^2}$$

x	$\varepsilon(x)$	δ	δ^2
$\vdots$	$\vdots$		
.04	$-.0001004$		
.05	1112		
.06	1177		
.08	1199	22	-111
.10	1110	-89	
.12	947		
$\vdots$	$\vdots$		

$$1199 - \frac{(22 - 89)^2}{8(-111)} = 1204$$

(19) Note: Our precise estimate of each d_i was obtained by taking the extremal ordinate of a polynomial of second degree passing through three equally spaced points on the error curve bracketing the peak.

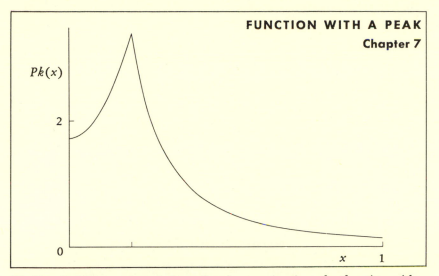

$Pk(x)$

2

0

x 1

(1) Our decision to attempt the rational approximation of a function with a peak was simply based on a desire to see what would happen. And, sparing but a few of the gory details, here is what occurred.

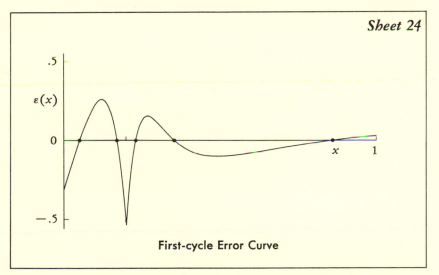

Sheet 24

.5

$\varepsilon(x)$

0

x 1

$-.5$

First-cycle Error Curve

(2) On the first cycle of the Sheet 24 case we began by imposing roots. We didn't have much of an idea as to where to put the roots, but we thought that the neighborhood of the peak would be quite critical. We placed two roots to the left of the peak and three to the right.

75

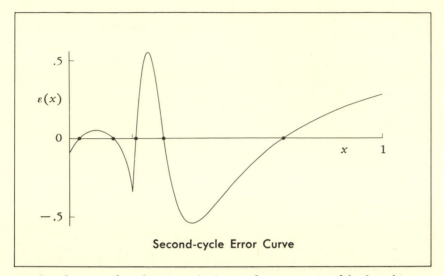

Second-cycle Error Curve

(3) On the second cycle we again imposed roots, as we felt that the case would be a pathological one in some respects, and hence the imposing of extremals might be a bit tricky. We thought that we were trying more reasonable locations, and therefore the above result was a bit of a jolt.

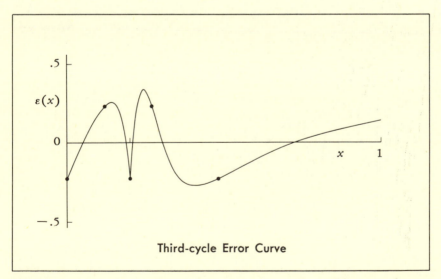

Third-cycle Error Curve

(4) However, the two cycles of data gave us enough confidence to dare to impose extremal deviations of guessed-at magnitude on the third cycle. The result was quite pleasing, and the case fell several cycles later without teaching us much.

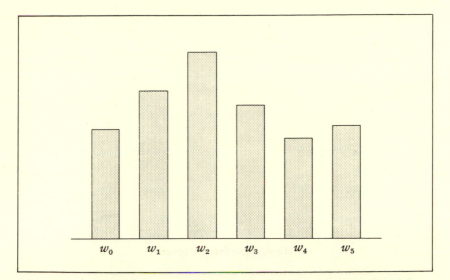

(5) Long after the three approximations discussed in this chapter had been leveled, we perturbed the final Sheet 24 approximation and obtained the above weights. Notice that the heaviest weight is associated with the peak deviation.

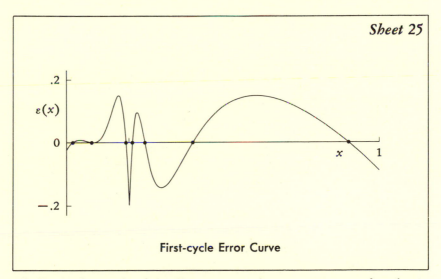

First-cycle Error Curve

(6) On the first cycle of the Sheet 25 case, the error curve went haywire on the left end. This led us to believe that maybe we should have imposed two roots to the left of the peak and five to the right, instead of doing what we did.

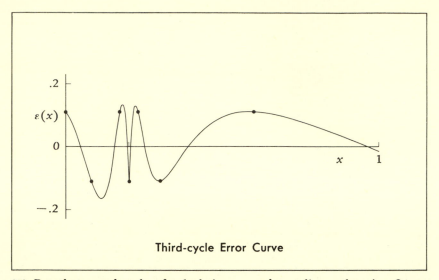

Third-cycle Error Curve

(7) But the second cycle of calculation seemed to tell us otherwise. So we took another look at the first-cycle error curve and decided that maybe the trouble would go away if we just ignored it. And, strangely enough, it did.

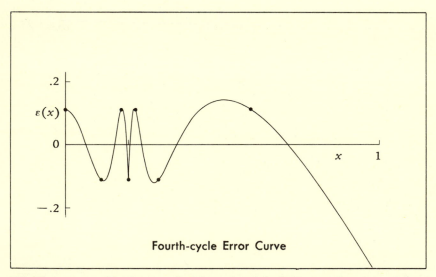

Fourth-cycle Error Curve

(8) Now up to this time we had been leaving the extremal at $x = 1$ free, thinking that it should be rather heavily weighted. We now learned that this end extremal was not at all stable and began to suspect that the peak at $x = .2$ was the dominant one.

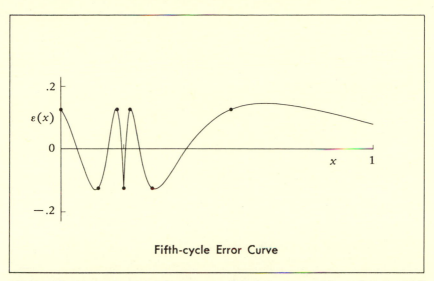

Fifth-cycle Error Curve

(9) But, for reasons now unknown, we failed to act on this conjecture until another cycle of calculation had been completed.

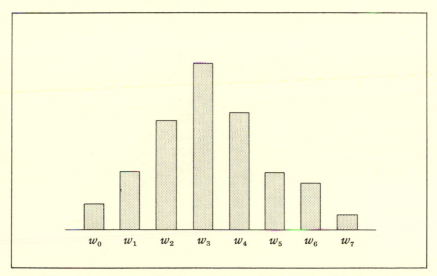

(10) We now put the reader in a position of "one-upness" before the page is turned. These are the weights associated with the approximation of Sheet 25. They were determined long after the three approximations discussed here were derived.

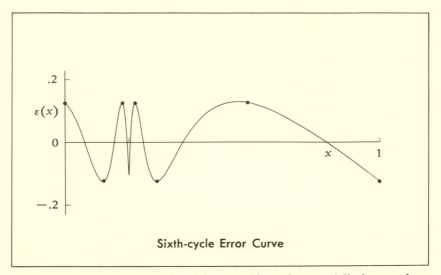

Sixth-cycle Error Curve

(11) We began to wonder if weights should not be especially heavy where roots crowd up to overcome a difficulty, and we decided to make up some weights that would go gently uphill in both directions to the number four, or peak, extremal.

Sixth Cycle		Seventh Cycle	
125	2	121	4
126	3	122	5
125	4	122	8
106	5	116	15
125	4	122	8
125	3	121	5
128	2	121	3
125	1	121	2
121	24	120	50

(12) And so the sixth-cycle deviations were cautiously weighted, as shown. When the seventh-cycle deviations were obtained, lumping indicated that perhaps one-third of the total weight should be assigned to the peak at $x = .2$. Perhaps the weights should look something like this, we thought.

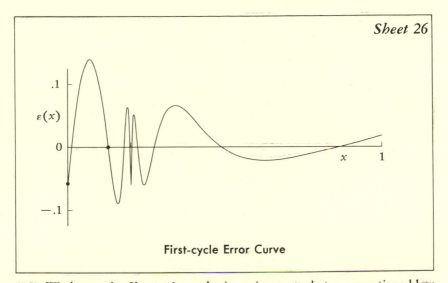

First-cycle Error Curve

Sheet 26

(13) We began the Sheet 26 case by imposing roots, but our equations blew up. We managed to redeem ourselves by throwing away the equation for smallest x and making up for this deletion by guessing at a value for a_0. (This is equivalent to imposing an extremal deviation at $x = 0$.)

First Cycle		Second Cycle		Third Cycle
.06	4	640	4	663
.14	5	657	5	663
.09	6	753	6	663
.06	9	662	9	666
.06	15	629	15	661
.05	9	642	9	660
.06	6	644	6	661
.07	5	696	5	663
.02	4	745	4	663
.02	3	640	3	663
.064	66	.0663	66	

(14) We then dreamed up what we thought to be a reasonable set of weights and applied them to the first-cycle deviations. These proved so effective that we re-used them on the second cycle and the case fell on the third. Thus ended our struggles with the $Pk(x)$ function.

Sheet

1 $\quad c_1\left(\dfrac{x-1}{x+1}\right) + c_3\left(\dfrac{x-1}{x+1}\right)^3$

2 $\quad c_1\left(\dfrac{x-1}{x+1}\right) + c_3\left(\dfrac{x-1}{x+1}\right)^3 + c_5\left(\dfrac{x-1}{x+1}\right)^5$

3 $\quad c_1\left(\dfrac{x-1}{x+1}\right) + c_3\left(\dfrac{x-1}{x+1}\right)^3 + c_5\left(\dfrac{x-1}{x+1}\right)^5 + c_7\left(\dfrac{x-1}{x+1}\right)^7$

4 $\quad c_1\left(\dfrac{x-1}{x+1}\right) + c_3\left(\dfrac{x-1}{x+1}\right)^3 + c_5\left(\dfrac{x-1}{x+1}\right)^5 + c_7\left(\dfrac{x-1}{x+1}\right)^7 + c_9\left(\dfrac{x-1}{x+1}\right)^9$

(1) We shall again use the letter ε to denote the greatest error of approximation; and, as we consider mostly best approximation, it will usually be true that $\varepsilon = d$. In this chapter, we consider ε as a function of parametric form and begin by examining Sheets 1, 2, 3, and 4.

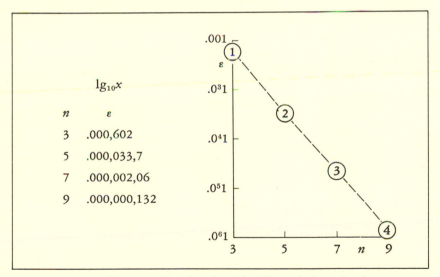

$\lg_{10} x$

n	ε
3	.000,602
5	.000,033,7
7	.000,002,06
9	.000,000,132

(2) A linear variable n is introduced to identify each form in the sequence, and ε is plotted against n on semilogarithmic paper. A nearly linear relationship is the result. Such linearity appears to be the rule rather than the exception, in our work.

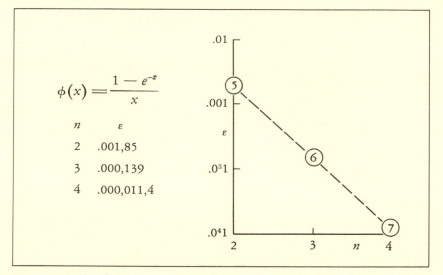

(3) In a similar fashion, we present here a diagram for the $\phi(x)$ approximations of Sheets 5, 6, and 7. Notice that the error is only decreasing by about
a factor of ten for each two parameters added. This is rather slow.

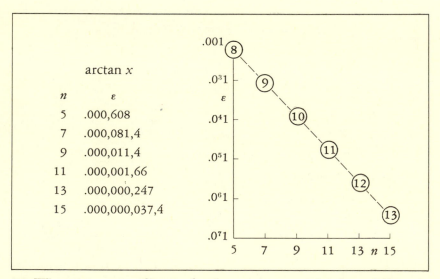

(4) We next present a diagram for the arctan x approximations of Sheets 8,
9, 10, 11, 12, and 13. Here the error decreases by about a factor of seven for
each parameter added. This is fairly good, to our way of thinking.

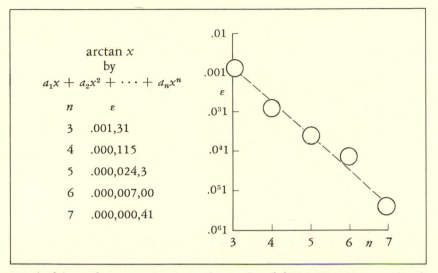

arctan x
by
$a_1x + a_2x^2 + \cdots + a_nx^n$

n	ε
3	.001,31
4	.000,115
5	.000,024,3
6	.000,007,00
7	.000,000,41

(5) And here, for comparison, are the results of fitting arctan x over $(0, 1)$ by the polynomial form without gaps save for the constant term a_0.

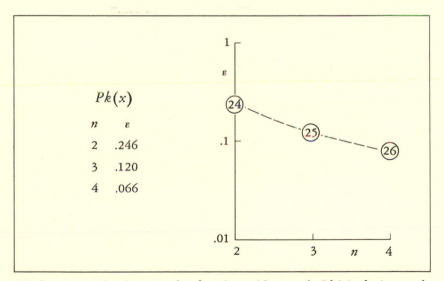

$Pk(x)$

n	ε
2	.246
3	.120
4	.066

(6) Our approximations to the function with a peak $Pk(x)$ don't exactly converge in a spectacular fashion. This is to be expected in the rational approximation of functions with derivatives that are discontinuous or infinite at one or more points.

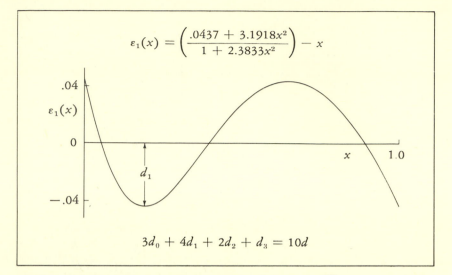

$$\varepsilon_1(x) = \left(\frac{.0437 + 3.1918x^2}{1 + 2.3833x^2} \right) - x$$

$$3d_0 + 4d_1 + 2d_2 + d_3 = 10d$$

(7) For another sad example, consider the fitting of $\sqrt{x}$ over $(0, 1)$ by a polynomial of nth degree in x over a polynomial of nth degree in x, or the equivalent problem of fitting x by the same parametric forms as functions of x^2.

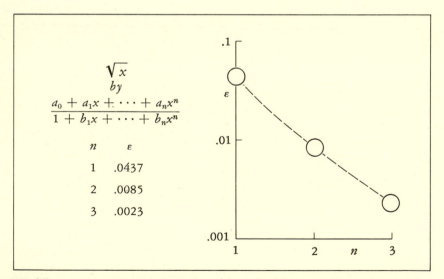

$$\sqrt{x}$$
by
$$\frac{a_0 + a_1x + \cdots + a_nx^n}{1 + b_1x + \cdots + b_nx^n}$$

n	ε
1	.0437
2	.0085
3	.0023

(8) These are our empirical results of best fit. The convergence is not what might be called rapid, nor should we expect it to be. But how fast do the points sink with increasing n?

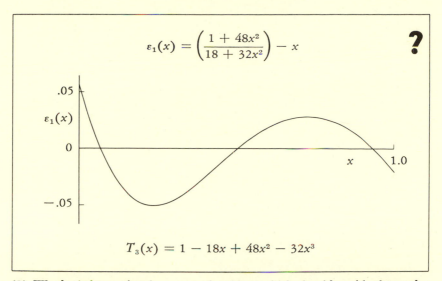

$$\varepsilon_1(x) = \left(\frac{1 + 48x^2}{18 + 32x^2}\right) - x$$

?

$$T_3(x) = 1 - 18x + 48x^2 - 32x^3$$

(9) We don't know, but here are a few hints which should enable the reader to construct an infinite sequence of rational approximations to $\sqrt{x}$ over $(0, 1)$ in which the convergence of ε_n is like A/n^2. Can a better result be obtained?

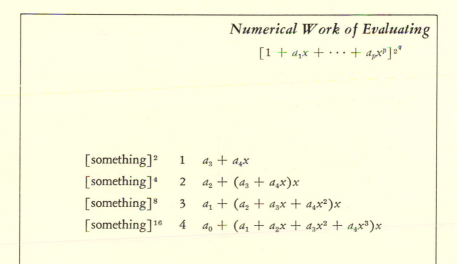

Numerical Work of Evaluating

$$[1 + a_1x + \cdots + a_px^p]^{2^q}$$

[something]^2	1	$a_3 + a_4x$
[something]^4	2	$a_2 + (a_3 + a_4x)x$
[something]^8	3	$a_1 + (a_2 + a_3x + a_4x^2)x$
[something]^{16}	4	$a_0 + (a_1 + a_2x + a_3x^2 + a_4x^3)x$

(10) We now describe three studies pertaining to the approximation of exponential and exponential-like functions. We begin by pointing out that q multiplications are required to raise a quantity to the 2^qth power, and p multiplications are required to evaluate a polynomial of pth degree.

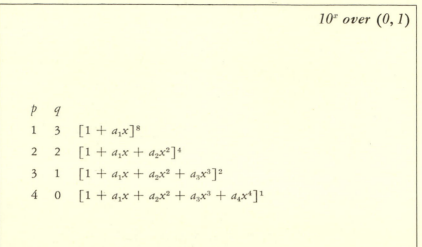

10^x *over* $(0, 1)$

p	q	
1	3	$[1 + a_1 x]^8$
2	2	$[1 + a_1 x + a_2 x^2]^4$
3	1	$[1 + a_1 x + a_2 x^2 + a_3 x^3]^2$
4	0	$[1 + a_1 x + a_2 x^2 + a_3 x^3 + a_4 x^4]^1$

(11) And so the reader may verify by inspection that each of the parametric forms listed above requires four multiplications to evaluate. In our first study, each of these forms was best-fitted to 10^x over $(0, 1)$ in the sense of minimum relative error.

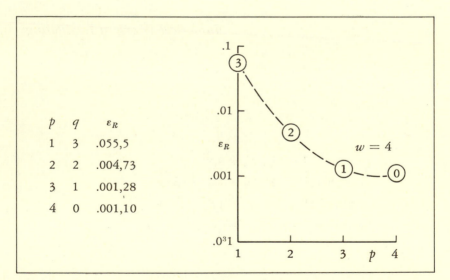

p	q	ε_R
1	3	.055,5
2	2	.004,73
3	1	.001,28
4	0	.001,10

(12) The points thus obtained were then joined together by a dashed line. The result is one connected string of points in our final chart. We shall say that the work of evaluation is $w = 4$ (multiplications) for these cases.

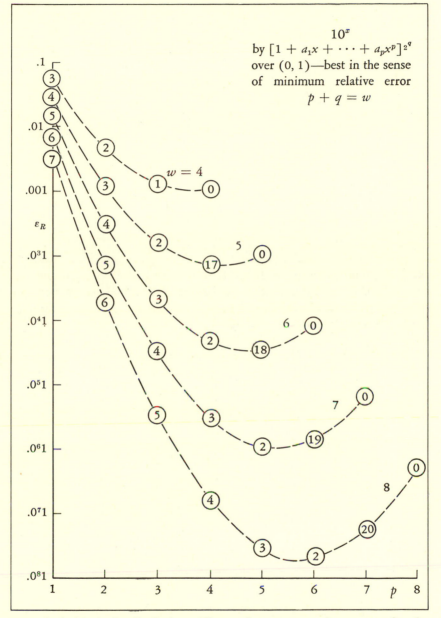

$$10^x$$
by $[1 + a_1 x + \cdots + a_p x^p]^{2^q}$
over $(0, 1)$—best in the sense
of minimum relative error
$$p + q = w$$

(13) And this is the final chart. The reader may note that, except for the points corresponding to Sheets 17, 18, 19, and 20, which are so labeled, we have written the appropriate q value in each little circle representing a point.

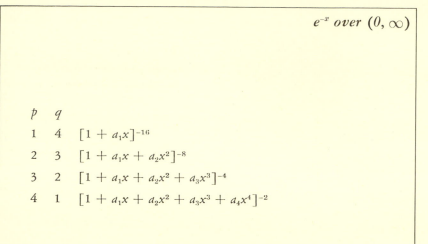

(14) Our second parameter study involved the fitting of e^{-x} over $(0, \infty)$ by reciprocals of the forms considered in the first study. Our sense of best fit is once again the usual one—that the greatest absolute error of approximation shall be made a minimum.

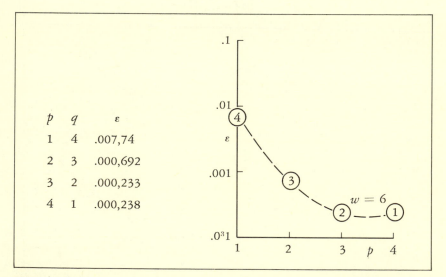

(15) Again we joined cases of equal work together in constructing the final diagram. The forms considered on this page all require five multiplications plus one division to evaluate. We shall say that the work of evaluation is $w = 6$ for these cases.

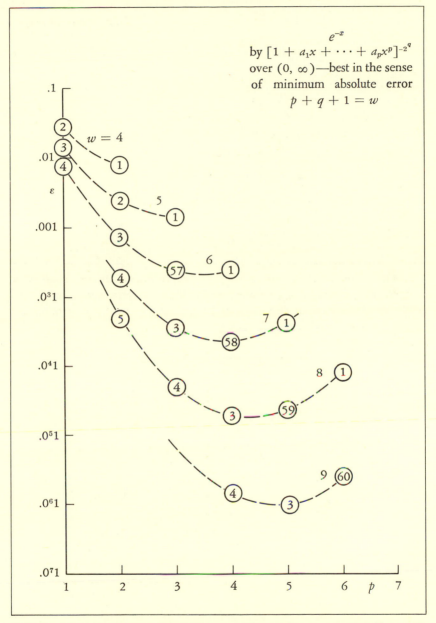

approximating e^{-x} by $[1 + a_1 x + \cdots + a_p x^p]^{-2^q}$ over $(0, \infty)$—best in the sense of minimum absolute error

$$p + q + 1 = w$$

(16) And here is the completed diagram. The reader can see that, for moderate accuracy of approximation, the fixed choice of a fourth-power exponent is not a bad one. The parametric forms studied here have been found very useful in approximating functions whose asymptotic behavior is like that of a decaying exponential as $x \to \infty$.

$\Phi(x)$ *over* $(0, \infty)$

p	q	
1	4	$1 - [1 + a_1 x]^{-16}$
2	3	$1 - [1 + a_1 x + a_2 x^2]^{-8}$
3	2	$1 - [1 + a_1 x + a_2 x^2 + a_3 x^3]^{-4}$
4	1	$1 - [1 + a_1 x + a_2 x^2 + a_3 x^3 + a_4 x^4]^{-2}$

(17) Our third parameter study involved the fitting of the Gaussian error integral $\Phi(x)$ over $(0, \infty)$ by unity minus the parametric forms considered in the second study. Our sense of best fit is again the usual one of minimum absolute error.

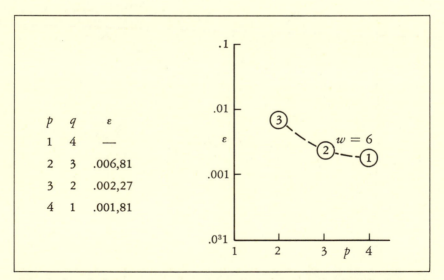

p	q	ε
1	4	—
2	3	.006,81
3	2	.002,27
4	1	.001,81

(18) And, as before, we joined cases involving equal work of evaluation. It is interesting to note in examining the final diagram that the strings of points for fixed q become quite irregular in rate of convergence as q increases.

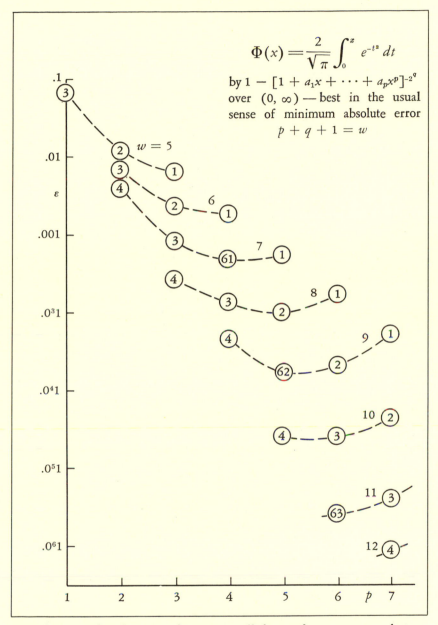

$$\Phi(x) = \frac{2}{\sqrt{\pi}} \int_0^x e^{-t^2}\, dt$$

by $1 - [1 + a_1 x + \cdots + a_p x^p]^{-2^q}$ over $(0, \infty)$ — best in the usual sense of minimum absolute error

$$p + q + 1 = w$$

$w = 5$

(19) Again, a fourth power does very well for moderate accuracy; but, as variety is the spice of life, we decided to present the three cases so marked as Sheets 61, 62, and 63. The sequence of parametric forms so used has a simple law of formation and gives rapidly increasing accuracy per step.

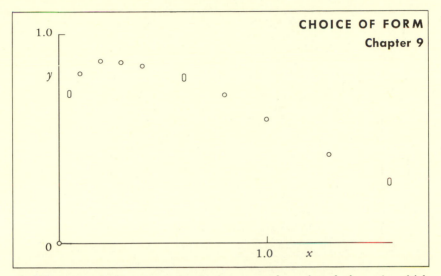

(1) We were asked to approximate the scanty data plotted above, in which the ovals denote especially untrustworthy points. The desired curve presumably starts at $(0, 0)$, rises steeply, follows the trend of the data, and then goes down to zero asymptotically as $x \to \infty$.

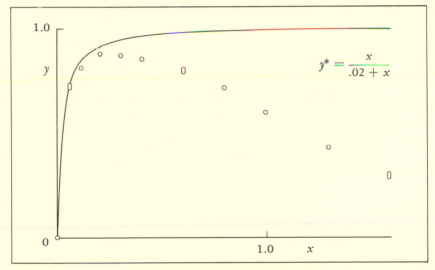

$$y^* = \frac{x}{.02 + x}$$

(2) We began by writing an initial approximation having about the right appearance in the neighborhood of the y-axis. Our plan was then to add whatever appeared necessary to the denominator to bring the curve down properly as x becomes large, but to leave it unaffected for small x.

95

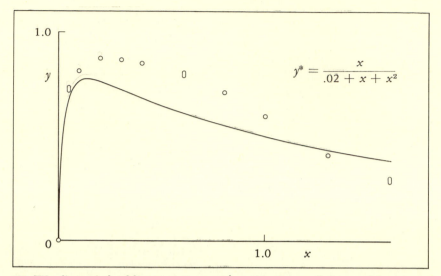

(3) We first tried adding an x^2 term, but the shape of the resulting curve doesn't look quite right, as you can see. The x^2 term adds too much downstairs for small x and not enough downstairs for large x.

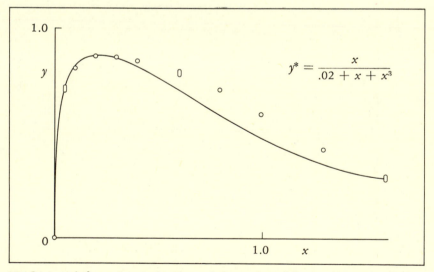

(4) So we tried an x^3 term in place of the x^2 term. The result appears to be a step in the right direction, and the shape of the curve is vastly improved. However, the x^3 term still appears to contribute too much downstairs for small x and not enough downstairs for large x.

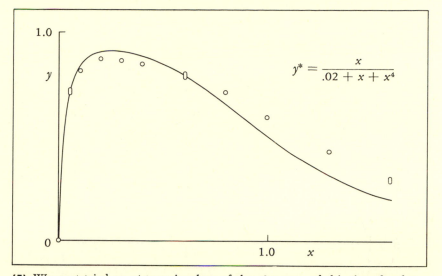

$$y^* = \frac{x}{.02 + x + x^4}$$

(5) We next tried an x^4 term in place of the x^3 term, and this time the shape of the curve appears to be about what it should be. Our next problem was that of adjusting the parameter values in the selected form to obtain a "best" fit.

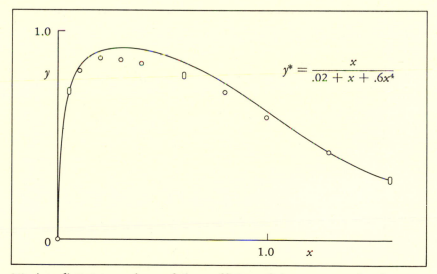

$$y^* = \frac{x}{.02 + x + .6x^4}$$

(6) As a first step, we lowered the coefficient of the x^4 term so that the curve now passes through the points on the far-right end.

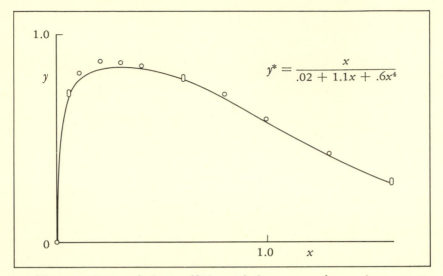

$$y^* = \frac{x}{.02 + 1.1x + .6x^4}$$

(7) Then we increased the coefficient of the x term downstairs so as to depress the curve in the middle range.

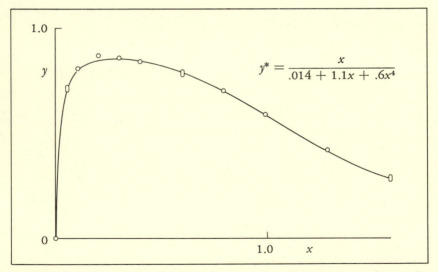

$$y^* = \frac{x}{.014 + 1.1x + .6x^4}$$

(8) We then lowered the constant term so as to raise the curve a bit on the left. The curve now appears to be a bit low in the neighborhood of its peak.

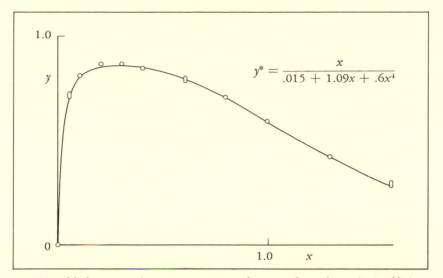

$$y^* = \frac{x}{.015 + 1.09x + .6x^4}$$

(9) We added .001 to the constant term and removed .01 from the coefficient of the x term, thus leaving the ordinate unchanged at $x = .1$. When the resulting curve was compared with the previous one, we concluded that another such step would do the job.

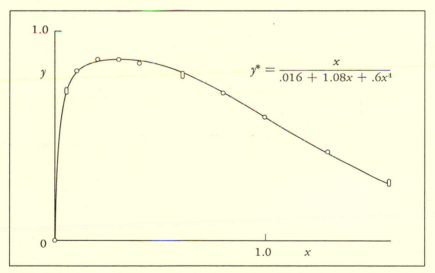

$$y^* = \frac{x}{.016 + 1.08x + .6x^4}$$

(10) And so we added another .001 to the constant term and removed another .01 from the coefficient of the x term downstairs. The resulting approximation is about as good as we can ask for, considering the shortcomings of the data supplied.

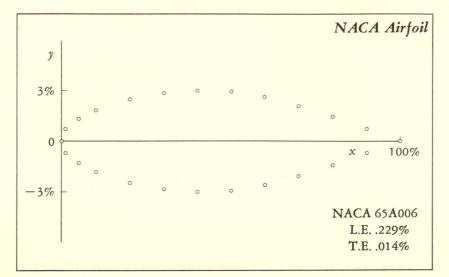

(11) As we were about to write this chapter, we were asked to suggest a parametric form that would be useful in the fitting of the wing section known as NACA 65A006. Our analysis of this problem follows.

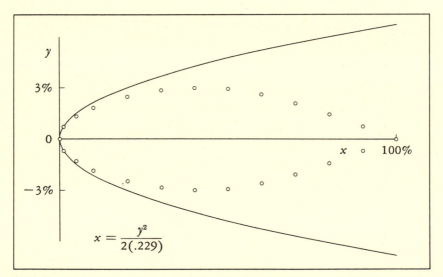

(12) As a first step, we determined a simple parabola $x = Ay^2$ that would fit a small region of the leading edge well. For this purpose we chose A so that the parabola would have the proper curvature at the origin.

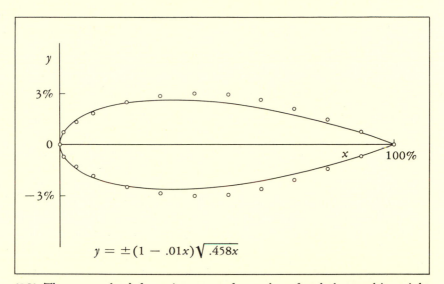

$$y = \pm(1 - .01x)\sqrt{.458x}$$

(13) Then we solved for y in terms of x and prefaced the resulting right member by a factor $(1 - .01x)$ to bring the curve down to zero at $x = 100\%$. As the trailing edge doesn't exactly come to a point, this is a slight violation of the data.

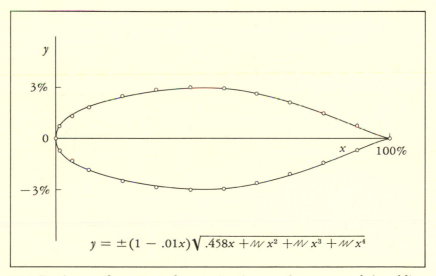

$$y = \pm(1 - .01x)\sqrt{.458x + \text{\it{w}} x^2 + \text{\it{w}} x^3 + \text{\it{w}} x^4}$$

(14) For increased accuracy of approximation, we then suggested the adding of further polynomial terms under the square root sign. The sequence of forms thus suggested proved to be quite satisfactory.

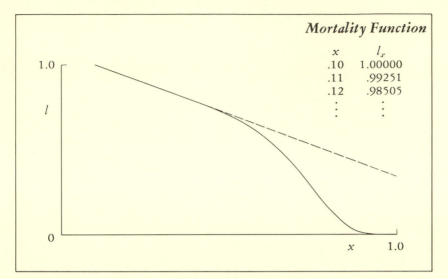

Mortality Function

x	l_x
.10	1.00000
.11	.99251
.12	.98505
$\vdots$	$\vdots$

(15) Our next example concerns the fitting of the familiar American Experience Mortality table, "decimalized" as indicated. The initial linearity of the curve and its subsequent nonlinear behavior argue against the success of low-order polynomial approximation.

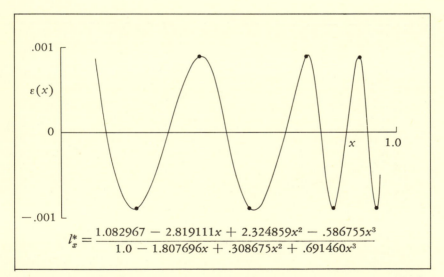

$$l_x^* = \frac{1.082967 - 2.819111x + 2.324859x^2 - .586755x^3}{1.0 - 1.807696x + .308675x^2 + .691460x^3}$$

(16) We attempted to fit a polynomial of second degree in x over a polynomial of second degree in x to l_x but were unable to level the error curve. Then we tried a cubic over cubic and obtained the rather remarkable approximation above.

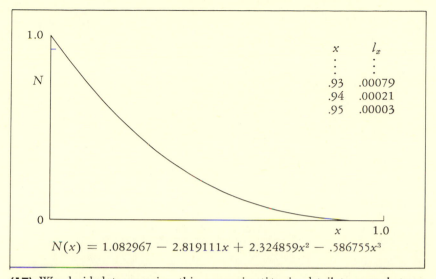

$$N(x) = 1.082967 - 2.819111x + 2.324859x^2 - .586755x^3$$

(17) We decided to examine this approximation in detail to see what we could learn. In round numbers, we discovered that $N(x)$ sinks to a minimum of $-.0001$ at $.94$ and has roots at $.93$, $.96$, and 2.1. Thus we seem to need a term like $(.94 - x)^2$ upstairs.

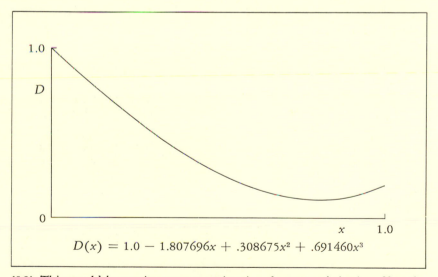

$$D(x) = 1.0 - 1.807696x + .308675x^2 + .691460x^3$$

(18) This would be to give our approximation the proper behavior off to the right. We also graphed $D(x)$, observed its parabolic appearance, and noted with interest how $N(x)$ and $D(x)$ conspire in quotient to fit the l_x curve so closely.

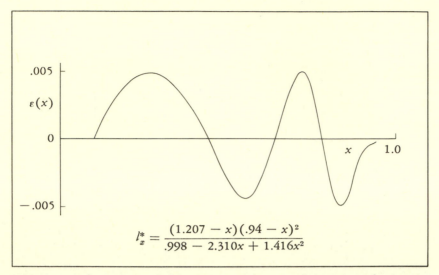

$$l_x^* = \frac{(1.207 - x)(.94 - x)^2}{.998 - 2.310x + 1.416x^2}$$

(19) A second requirement then appears to be a nearly equal quadratic term downstairs that can cancel the early effect of the $(.94 - x)^2$ term upstairs and yet not detract from the ability of this latter term to give the proper shape to the curve near the end of the table.

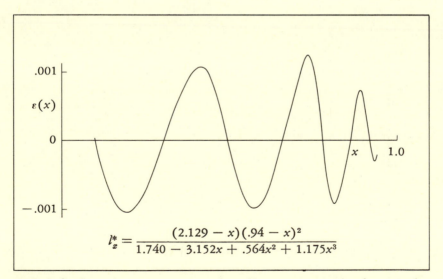

$$l_x^* = \frac{(2.129 - x)(.94 - x)^2}{1.740 - 3.152x + .564x^2 + 1.175x^3}$$

(20) A final requirement appears to be a linear term upstairs that can give the curve its required early linear appearance. Further terms then serve to supply added flexibility; but we may ask, at this point, "How really good are these data, anyway?"

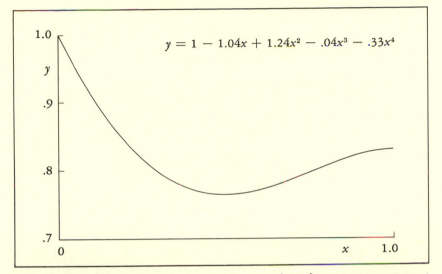

$$y = 1 - 1.04x + 1.24x^2 - .04x^3 - .33x^4$$

(21) Apart from knowing when something as awkward as a square root must be incorporated in a parametric form, one of the approximator's greatest problems is that of knowing when a low-order polynomial has a chance of doing a job and when it hasn't.

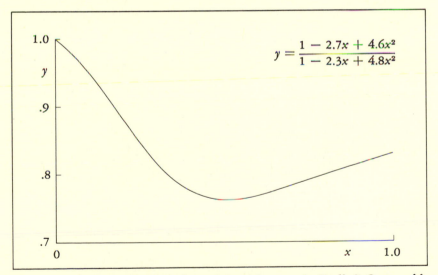

$$y = \frac{1 - 2.7x + 4.6x^2}{1 - 2.3x + 4.8x^2}$$

(22) Roughly speaking, low-order polynomials are quite "rolly." Comparable low-order rational expressions, however, can be quite "angular." The low-order rational expression can turn sharply and then go much straighter than the low-order polynomial can, with such a start.

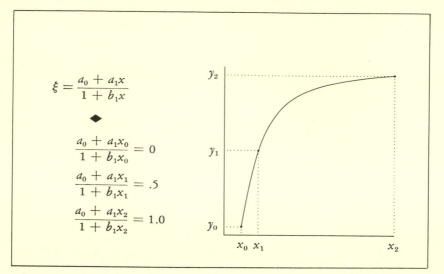

(23) Often we run into curves that may be made quite polynomial-like by the use of the simple rational transformation $x \to \xi$ defined above. To determine suitable parameter values, we divide the vertical range of variation into two (roughly) equal parts.

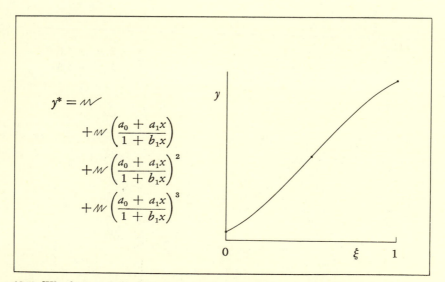

(24) We then require that x_0, x_1, and x_2 transform to equally spaced ξ values of, say, 0, .5, and 1.0. The transformed curve is then approximated by a polynomial in ξ in the usual fashion.

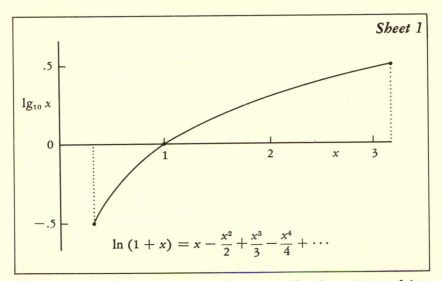

$$\ln(1+x) = x - \frac{x^2}{2} + \frac{x^3}{3} - \frac{x^4}{4} + \cdots$$

(25) In this connection, it is interesting to consider the sequence of logarithmic approximations, beginning with Sheet 1. Polynomials of interestingly low orders don't handle a full logarithmic cycle very well.

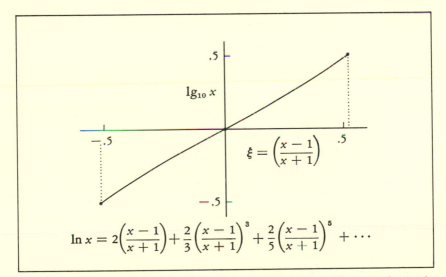

$$\ln x = 2\left(\frac{x-1}{x+1}\right) + \frac{2}{3}\left(\frac{x-1}{x+1}\right)^3 + \frac{2}{5}\left(\frac{x-1}{x+1}\right)^5 + \cdots$$

(26) But now let us introduce the transformation $\xi(x) = (x-1)/(x+1)$. When plotted against the new variable, ξ, the three points become equally spaced horizontally as well as vertically, and the curve is now quite flat.

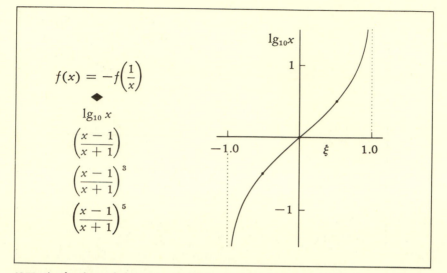

(27) And other advantages every bit as great also accrue. Thus $\xi(x)$ and any odd power of this quantity satisfy the functional relationship $f(x) = -f(x^{-1})$ satisfied by $\lg_{10} x$ itself. Thus the form indicated, if fitted over $(1, \sqrt{10})$, will automatically hold over the full logarithmic cycle $(1/\sqrt{10}, \sqrt{10})$.

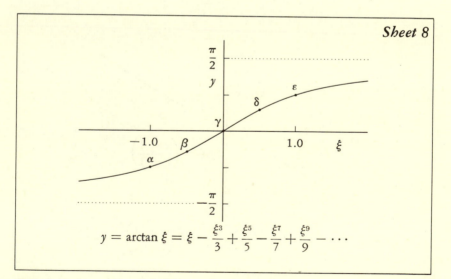

(28) Sometimes it is advisable merely to note that a function is even or odd in choosing a suitable parametric form. An approximation so made to hold over $(0, 1)$ then automatically holds over $(-1, 1)$.

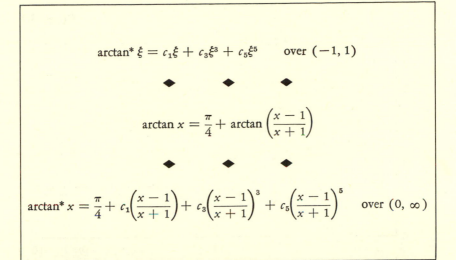

$$\text{arctan*}\ \xi = c_1\xi + c_3\xi^3 + c_5\xi^5 \qquad \text{over } (-1, 1)$$

$$\text{arctan } x = \frac{\pi}{4} + \arctan\left(\frac{x-1}{x+1}\right)$$

$$\text{arctan*}\ x = \frac{\pi}{4} + c_1\left(\frac{x-1}{x+1}\right) + c_3\left(\frac{x-1}{x+1}\right)^3 + c_5\left(\frac{x-1}{x+1}\right)^5 \qquad \text{over } (0, \infty)$$

(29) In the case of the arctan x function, an approximation holding over $(-1, 1)$ can be transformed into one holding over $(0, \infty)$ by using the addition law of this function in the indicated manner.

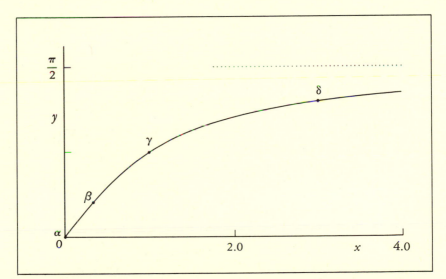

(30) And here, strangely enough, we have used exactly the same transformation to straighten out the arctan x function over $(0, \infty)$ as we did to straighten out the logarithmic function over $(1/\sqrt{10}, \sqrt{10})$.

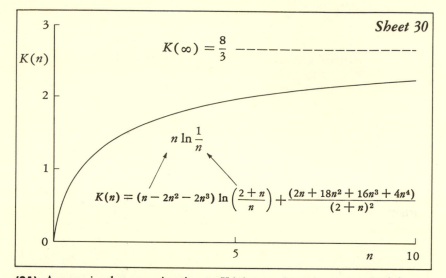

$$K(\infty) = \frac{8}{3}$$

$$n \ln \frac{1}{n}$$

$$K(n) = (n - 2n^2 - 2n^3) \ln\left(\frac{2+n}{n}\right) + \frac{(2n + 18n^2 + 16n^3 + 4n^4)}{(2+n)^2}$$

(31) A very simple approximation to $K(n)$ over $(0, \infty)$ was required for use on a high-speed digital computing machine. If at all possible, we were to obtain a rational fit, but observe: $K(n)$ has an infinite derivative at $n = 0$!

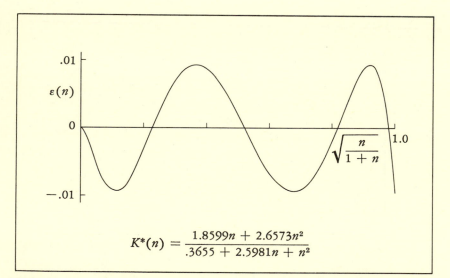

$$K^*(n) = \frac{1.8599n + 2.6573n^2}{.3655 + 2.5981n + n^2}$$

(32) And yet, despite the apparent difficulty of the task, a quadratic over a quadratic gave us a fit good to about .01, and a cubic over a cubic gave a fit good to about .001. Perhaps the indicated singularity is not quite so bad as a $\sqrt{n}$ singularity would be.

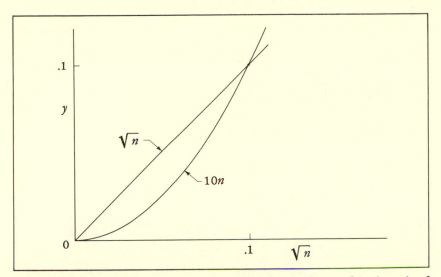

(33) This is easily seen to be the case, and the fact is illustrated in the pair of diagrams on this page. Against a $\sqrt{n}$ horizontal scale, the function $y = \sqrt{n}$ is a straight line with unit slope, whereas the function $y = -n \ln n$ is a curve with zero slope at the origin.

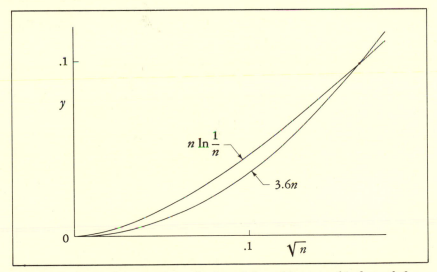

(34) Both the quadratic over quadratic and the cubic over cubic forms behave like An for small n, and against a $\sqrt{n}$ scale An plots as a parabola with zero slope at the origin.

Sheet 46

$$K(k) = \int_0^{\pi/2} \frac{d\phi}{\sqrt{1 - k^2 \sin^2 \phi}}$$

$$K(k) = \Lambda + \left(\frac{\Lambda - 1}{4}\right)k'^2 + \frac{9}{64}\left(\Lambda - \frac{7}{6}\right)k'^4 + \cdots$$

$$\Lambda = \ln\frac{4}{k'} \qquad k' = \sqrt{1 - k^2}$$

$$K^*(k) = [a_0 + a_1(1 - k^2) + a_2(1 - k^2)^2]$$

$$+ [b_0 + b_1(1 - k^2) + b_2(1 - k^2)^2] \ln\left(\frac{1}{1 - k^2}\right)$$

(35) In preparing the sheets of Part II, well-known expansions of one kind or another usually guided us in our choice of parametric form. Thus, the first few terms of a curious elliptic integral expansion about $k = 1$ had the above form when rewritten.

Sheet 64

$$-Ei(-x) = \int_x^\infty \frac{e^{-t}}{t}\,dt$$

$$-Ei(-x) \sim \frac{e^{-x}}{x}\left[1 - \frac{1!}{x} + \frac{2!}{x^2} - \frac{3!}{x^3} + \cdots\right]$$

$$-Ei(-x) = -\gamma - \ln x + x - \frac{x^2}{2 \cdot 2!} + \frac{x^3}{3 \cdot 3!} - \cdots$$

$$-Ei^*(-x) = \frac{e^{-x}}{x}\left[\frac{a_0 + a_1 x + x^2}{b_0 + b_1 x + x^2}\right]$$

(36) Matching the asymptotic properties of a function often allows us to obtain an expansion over an interval extending to ∞. In this case, the function has a logarithmic singularity at the origin and so we settled for approximation over $(1, \infty)$.

$$q = \frac{1}{\sqrt{2\pi}} \int_{x(q)}^{\infty} e^{-\frac{1}{2}t^2} \, dt$$

$$q \sim \frac{e^{-\frac{1}{2}x^2}}{\sqrt{2\pi}} \left[\frac{1}{x} - \frac{1}{x^3} + \frac{1 \cdot 3}{x^5} - \cdots \right]$$

$$q = e^{-\frac{1}{2}\eta^2} \to \eta = \sqrt{\ln \frac{1}{q^2}}$$

$$x^*(q) = \eta - \left[\frac{a_0 + a_1 \eta}{1 + b_1 \eta + b_2 \eta^2} \right]$$

(37) In the case of the inverse Gaussian, we examined the asymptotic expression and then introduced the new variable η. We could then see that x was equal to η minus a quantity that $\to 0$ as $\eta \to \infty$. Hence our choice of form.

$$w(z) = \int_0^{\infty} \frac{e^{-uz}}{K_1^2(u) + \pi^2 I_1^2(u)} \frac{du}{u}$$

$$w(z) = \frac{1}{2} - \frac{3}{8}z + \frac{3}{16}z^2 - \cdots$$

$$w(z) \sim \frac{1}{z^2} + \frac{6}{z^4} \left(\ln 2z - \frac{4}{3} \right) + \cdots$$

$$w^*(z) = \frac{1 + a_1 z}{2 + b_1 z + b_2 z^2 + b_3 z^3}$$

(38) In this case, the integral expression tells us that $w(z)$ is monotone decreasing over $(0, \infty)$. The asymptotic expansion tells us that $w(z)$ goes to zero like a constant over z^2. Our choice of form for approximation over $(0, \infty)$ then seems quite obvious.

REFERENCES

1. ABBOTT, IRA H., AND ALBERT E. VON DOENHOFF, *Theory of Wing Sections*, McGraw-Hill Book Company, Inc., New York, 1949.
2. HODGMAN, CHARLES D., *Mathematical Tables*, Chemical Rubber Publishing Company, Cleveland, Ohio, 1947.

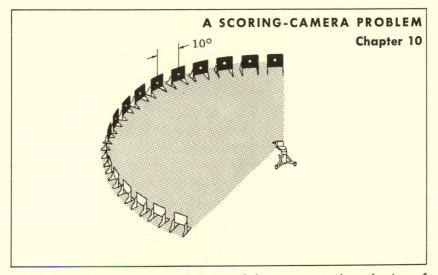

(1) Our concluding example of choice of form concerns the reduction of scoring-camera data: the problem to be discussed was referred to us by Mr. John Lowe, of the Douglas Aircraft Company.

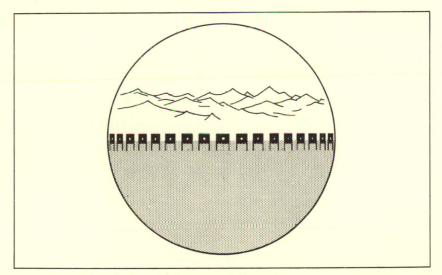

(2) To calibrate a scoring camera, a target array is set up and is photographed by the camera to be calibrated. The optical axis of the lens system is centered on the middle target of the array.

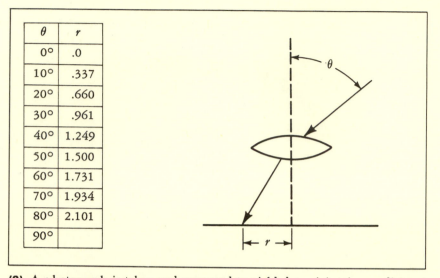

θ	r
0°	.0
10°	.337
20°	.660
30°	.961
40°	1.249
50°	1.500
60°	1.731
70°	1.934
80°	2.101
90°	

(3) A photograph is taken and measured to yield data giving image distance r as a function of entering-ray angle θ. The distance r is measured from the point on the film intersected by the optical axis. A sample of such calibrating data was given to us by Mr. Lowe.

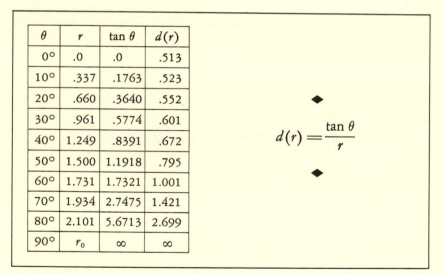

θ	r	$\tan \theta$	$d(r)$
0°	.0	.0	.513
10°	.337	.1763	.523
20°	.660	.3640	.552
30°	.961	.5774	.601
40°	1.249	.8391	.672
50°	1.500	1.1918	.795
60°	1.731	1.7321	1.001
70°	1.934	2.7475	1.421
80°	2.101	5.6713	2.699
90°	r_0	∞	∞

$$d(r) = \frac{\tan \theta}{r}$$

(4) And we were asked to suggest a simple parametric form that would be useful in the approximation of $d(r) = r^{-1} \tan \theta$. We began our investigation by getting out a table of tangents and computing a column of $d(r)$ values.

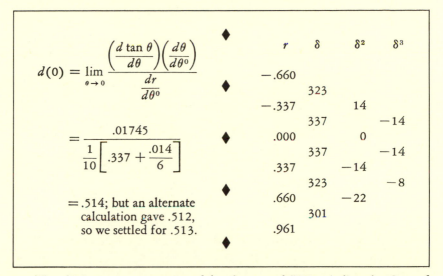

$$d(0) = \lim_{\theta \to 0} \frac{\left(\dfrac{d \tan \theta}{d\theta}\right)\left(\dfrac{d\theta}{d\theta^\circ}\right)}{\dfrac{dr}{d\theta^\circ}}$$

$$= \frac{.01745}{\dfrac{1}{10}\left[.337 + \dfrac{.014}{6}\right]}$$

$= .514$; but an alternate calculation gave $.512$, so we settled for $.513$.

r	δ	δ^2	δ^3
$-.660$			
	323		
$-.337$		14	
	337		-14
$.000$		0	
	337		-14
$.337$		-14	
	323		-8
$.660$		-22	
	301		
$.961$			

(5) The $d(0)$ entry was computed by the use of L'Hospital's rule. One of the derivatives involved had to be computed numerically, and for this purpose a familiar central difference formula was employed.

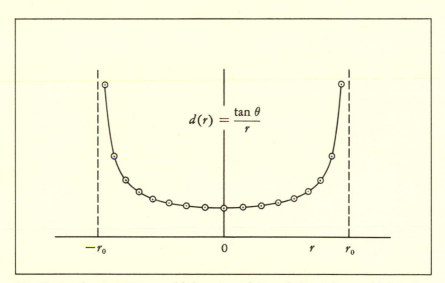

$$d(r) = \frac{\tan \theta}{r}$$

(6) Notice that $\tan \theta$ is an odd function of θ, and that θ is an odd function of r. As a result, $\tan \theta$ is an odd function of r, and so $d(r)$ must be an even function of r. If the $r(\theta)$ relationship is extrapolated smoothly to $r(90^\circ) = r_0$, then $d(r)$ will have simple poles at $\pm r_0$.

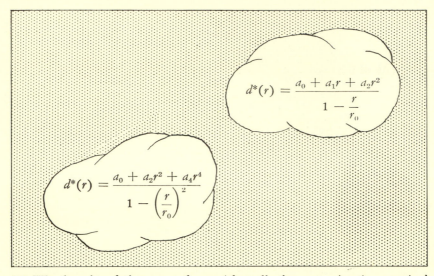

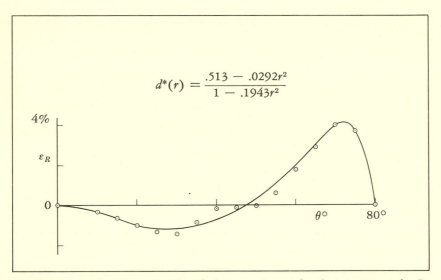

(7) We thought of the upper form. After all, the approximation required would only be used for positive r. But no, $d(r)$ is an even function of r, and it would only be good craftsmanship to say so. We thought of the lower form and felt much happier.

$$d^*(r) = \frac{.513 - .0292r^2}{1 - .1943r^2}$$

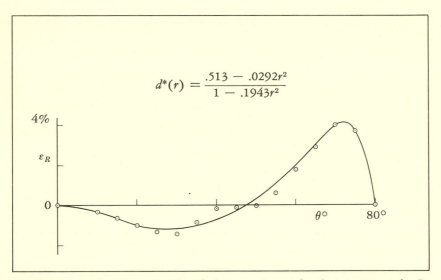

(8) As r_0 is unknown, we replaced the quantity r_0^{-2} by the parameter b_2. In our first attempt at fitting, we tried just two terms upstairs. To obtain additional points on the error curve, we interpolated midvalues of $r(\theta)$ and computed corresponding $d(r)$ values.

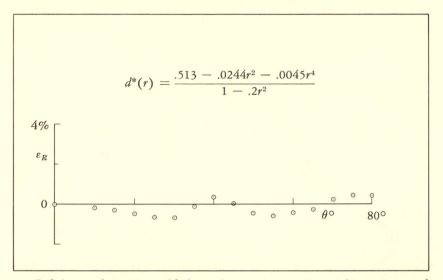

$$d^*(r) = \frac{.513 - .0244r^2 - .0045r^4}{1 - .2r^2}$$

(9) Judging as best we could from the previous case, we then set up and solved for the improved approximation above with three terms upstairs. An interesting but quite unimportant question came to mind.

θ°	r	δ	δ^2
50°	1.500		
		231	
60°	1.731		−28
		203	
70°	1.934		−36
		167	
80°	2.101		−32
		135	
90°	2.236		

(10) The value of b_2 appeared quite stable. Did it give us a reasonable estimate of r_0? To answer this question, we then computed $r_0 = \sqrt{5} = 2.236$ and formed a difference table. See how this number fits in the table above!

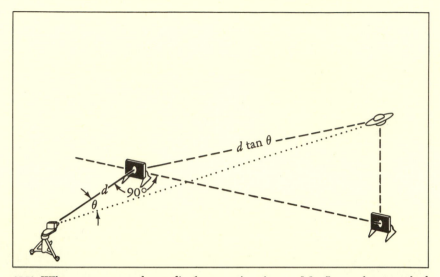

(11) When we reported our final approximation to Mr. Lowe, he remarked "You've come up with a function of r^2. That means I won't have to compute any square roots." It was only then that we thought to find out how the approximation was to be used.

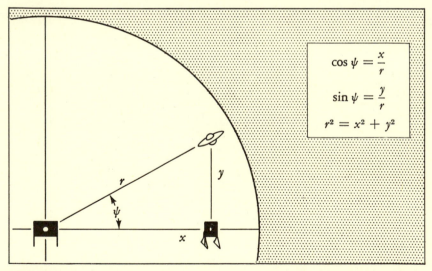

$$\cos \psi = \frac{x}{r}$$

$$\sin \psi = \frac{y}{r}$$

$$r^2 = x^2 + y^2$$

(12) And this is the remainder of the story: Coordinates x and y of the image of an object in space are read from the film. To clarify the geometry, we consider the space object to be vertically above the ground target on the right.

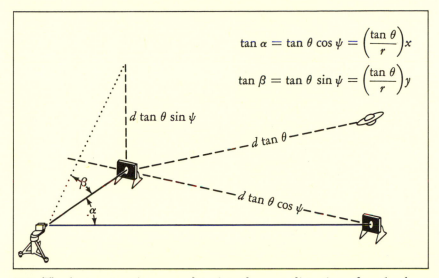

$$\tan \alpha = \tan \theta \cos \psi = \left(\frac{\tan \theta}{r}\right)x$$

$$\tan \beta = \tan \theta \sin \psi = \left(\frac{\tan \theta}{r}\right)y$$

(13) The lens system is assumed to introduce no distortion of angle about the optical axis. The big problem is that of converting film coordinates x and y into angular space coordinates α and β.

$$\tan \alpha \doteq \left[\frac{.513 - .0244(x^2 + y^2) - .0045(x^2 + y^2)^2}{1 - .2(x^2 + y^2)}\right]x$$

$$\tan \beta \doteq \left[\frac{.513 - .0244(x^2 + y^2) - .0045(x^2 + y^2)^2}{1 - .2(x^2 + y^2)}\right]y$$

(14) And so, by working out the little geometry and trigonometry involved, we learned why our equations would be used in the manner indicated above. Now, let's browse through the second part of the book!

PART II

Function:

$$\log_{10} X$$

Range:

$$\frac{1}{\sqrt{10}} \leq X \leq \sqrt{10}$$

Approximation:

$$\log^*_{10} X = C_1\left(\frac{X-1}{X+1}\right) + C_3\left(\frac{X-1}{X+1}\right)^3$$

$$C_1 = .86304$$

$$C_3 = .36415$$

Error Curve (Approximation–Function):

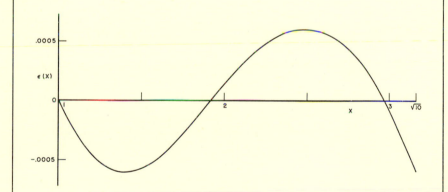

Comments: An equally good approximation over $1 \leq X \leq 10$ is

$$\log^*_{10} X = \frac{1}{2} + C_1\left(\frac{X-\sqrt{10}}{X+\sqrt{10}}\right) + C_3\left(\frac{X-\sqrt{10}}{X+\sqrt{10}}\right)^3.$$

Function:

$$\log_{10} X$$

Range:

$$\frac{1}{\sqrt{10}} \leq X \leq \sqrt{10}$$

Approximation:

$$\log^*_{10} X = C_1\left(\frac{X-1}{X+1}\right) + C_3\left(\frac{X-1}{X+1}\right)^3 + C_5\left(\frac{X-1}{X+1}\right)^5$$

$$C_1 = .8690286$$

$$C_3 = .2773839$$

$$C_5 = .2543275$$

Error Curve (Approximation–Function):

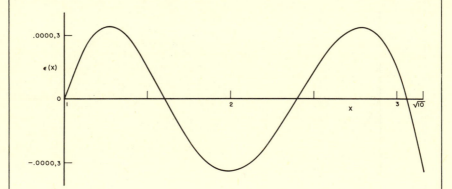

Comments: An equally good approximation over $1 \leq X \leq 10$ is

$$\log^*_{10} X = \frac{1}{2} + C_1\left(\frac{X-\sqrt{10}}{X+\sqrt{10}}\right) + C_3\left(\frac{X-\sqrt{10}}{X+\sqrt{10}}\right)^3 + C_5\left(\frac{X-\sqrt{10}}{X+\sqrt{10}}\right)^5.$$

Sheet: 3

Function:

$$\log_{10} X$$

Range:

$$\frac{1}{\sqrt{10}} \leq X \leq \sqrt{10}$$

Approximation:

$$\log_{10}^{*} X = C_1\left(\frac{X-1}{X+1}\right) + C_3\left(\frac{X-1}{X+1}\right)^3 + C_5\left(\frac{X-1}{X+1}\right)^5 + C_7\left(\frac{X-1}{X+1}\right)^7$$

$$C_1 = .8685, 5434 \qquad C_5 = .1536, 1371$$

$$C_3 = .2911, 5068 \qquad C_7 = .2113, 9497$$

Error Curve (Approximation–Function):

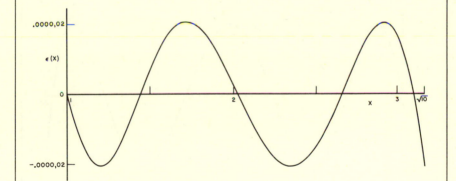

Comments: An equally good approximation over $1 \leq X \leq 10$ is

$$\log_{10}^{*} X = \frac{1}{2} + C_1\left(\frac{X-\sqrt{10}}{X+\sqrt{10}}\right) + C_3\left(\frac{X-\sqrt{10}}{X+\sqrt{10}}\right)^3 + \cdots + C_7\left(\frac{X-\sqrt{10}}{X+\sqrt{10}}\right)^7.$$

Function:

$$\log_{10} X$$

Range:

$$\frac{1}{\sqrt{10}} \leq X \leq \sqrt{10}$$

Approximation:

$$\log^*_{10}X = C_1\left(\frac{X-1}{X+1}\right) + C_3\left(\frac{X-1}{X+1}\right)^3 + \cdots + C_9\left(\frac{X-1}{X+1}\right)^9$$

$$C_1 = .8685,91718 \qquad C_7 = .0943,76476$$

$$C_3 = .2893,35524 \qquad C_9 = .1913,37714$$

$$C_5 = .1775,22071$$

Error Curve (Approximation–Function):

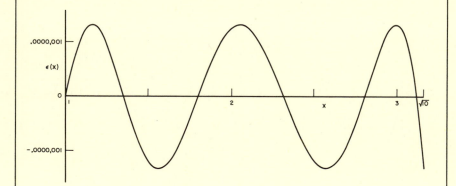

Comments: An equally good approximation over $1 \leq X \leq 10$ is

$$\log^*_{10}X = \frac{1}{2} + C_1\left(\frac{X-\sqrt{10}}{X+\sqrt{10}}\right) + C_3\left(\frac{X-\sqrt{10}}{X+\sqrt{10}}\right)^3 + \cdots + C_9\left(\frac{X-\sqrt{10}}{X+\sqrt{10}}\right)^9.$$

Sheet: 5

Function:

$$\varphi(X) = \frac{1 - e^{-X}}{X}$$

Range:

$$0 \le X < \infty$$

Approximation:

$$\xi = \frac{1}{1 + pX}$$

$$\varphi^*(X) = \frac{a_1\xi + a_2\xi^2}{1 + b_1\xi + b_2\xi^2}$$

$p = .47698$ $\quad a_1 = .42850$ $\quad b_1 = -.57953$

$\quad a_2 = .56965$ $\quad b_2 = .57953$

Error Curve (Approximation–Function):

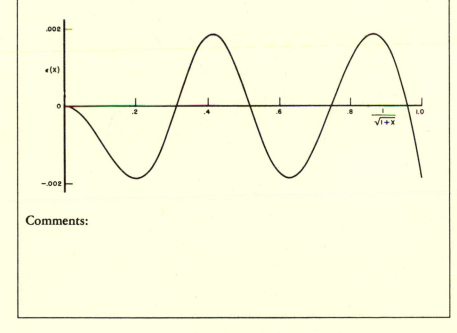

Comments:

Function:

$$\varphi(X) = \frac{1 - e^{-X}}{X}$$

Range:

$$0 \leq X < \infty$$

Approximation:

$$\xi = \frac{1}{1 + pX}$$

$$\varphi^*(X) = \frac{a_1\xi + a_2\xi^2 + a_3\xi^3}{1 + b_1\xi + b_2\xi^2 + b_3\xi^3}$$

$p = .3606032$ $a_1 = .3671626$ $b_1 = -1.3562710$

$a_2 = -.2272232$ $b_2 = 1.6148087$

$a_3 = .8601996$ $b_3 = -.2585377$

Error Curve (Approximation–Function):

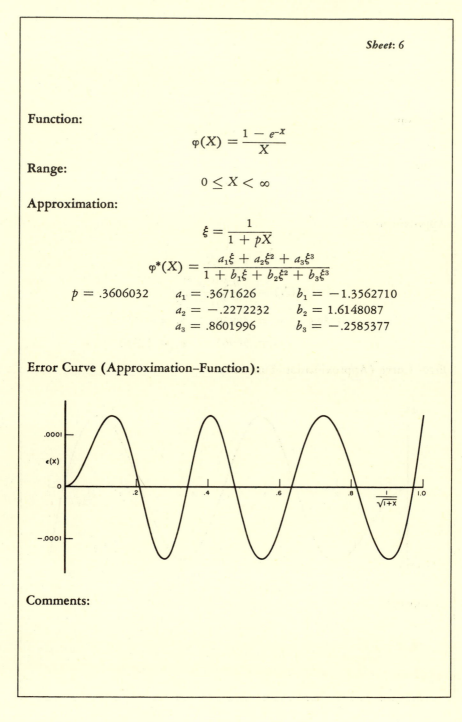

Comments:

Function:

$$\varphi(X) = \frac{1 - e^{-X}}{X}$$

Range:

$$0 \leq X < \infty$$

Approximation:

$$\xi = \frac{1}{1 + pX}$$

$$\varphi^*(X) = \frac{a_1\xi + a_2\xi^2 + a_3\xi^3 + a_4\xi^4}{1 + b_1\xi + b_2\xi^2 + b_3\xi^3 + b_4\xi^4}$$

$p = .2898,9933$ $\quad a_1 = .2890,5386$ $\quad b_1 = -2.2178,1431$

$\quad\quad\quad\quad\quad\quad a_2 = -.3324,0494$ $\quad b_2 = 3.3313,1912$

$\quad\quad\quad\quad\quad\quad a_3 = .4554,8498$ $\quad b_3 = -1.6278,1495$

$\quad\quad\quad\quad\quad\quad a_4 = .5878,5466$ $\quad b_4 = .5143,1014$

Error Curve (Approximation–Function):

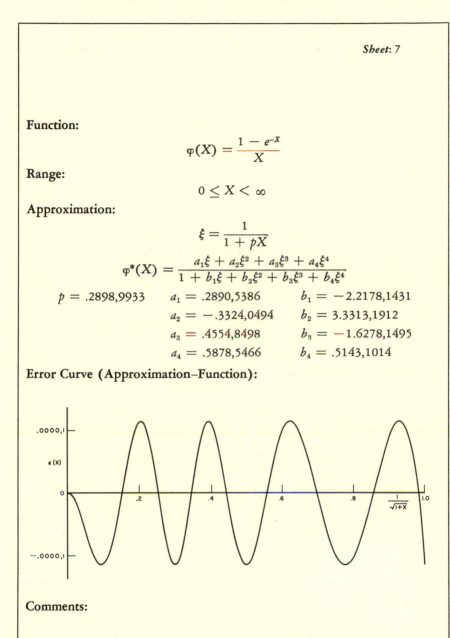

Comments:

Function:

$$\arctan X$$

Range:

$$-1 \leq X \leq 1$$

Approximation:

$$\arctan^* X = C_1 X + C_3 X^3 + C_5 X^5$$
$$C_1 = .995354$$
$$C_3 = -.288679$$
$$C_5 = .079331$$

Error Curve (Approximation–Function):

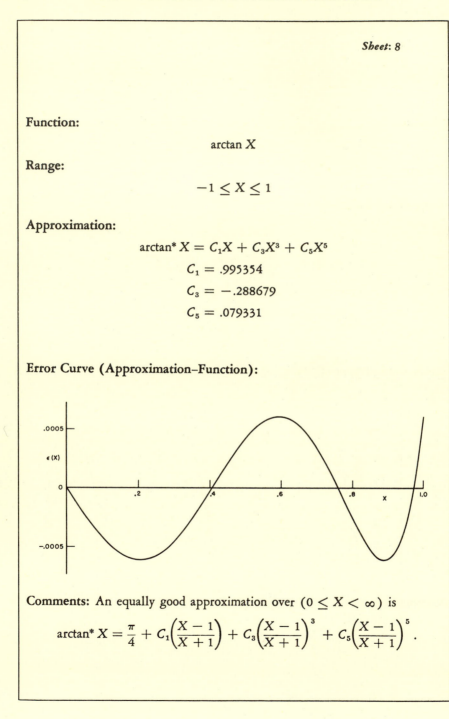

Comments: An equally good approximation over $(0 \leq X < \infty)$ is

$$\arctan^* X = \frac{\pi}{4} + C_1\left(\frac{X-1}{X+1}\right) + C_3\left(\frac{X-1}{X+1}\right)^3 + C_5\left(\frac{X-1}{X+1}\right)^5.$$

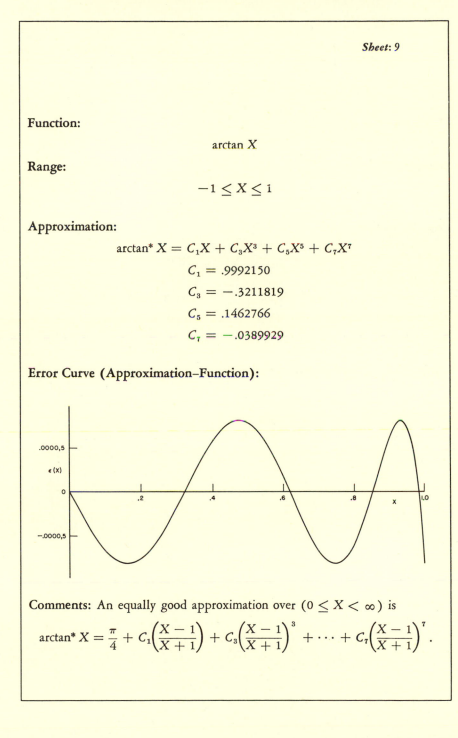

Sheet: 9

Function:

$$\arctan X$$

Range:

$$-1 \leq X \leq 1$$

Approximation:

$$\arctan^* X = C_1 X + C_3 X^3 + C_5 X^5 + C_7 X^7$$

$$C_1 = .9992150$$

$$C_3 = -.3211819$$

$$C_5 = .1462766$$

$$C_7 = -.0389929$$

Error Curve (Approximation–Function):

Comments: An equally good approximation over $(0 \leq X < \infty)$ is

$$\arctan^* X = \frac{\pi}{4} + C_1\left(\frac{X-1}{X+1}\right) + C_3\left(\frac{X-1}{X+1}\right)^3 + \cdots + C_7\left(\frac{X-1}{X+1}\right)^7.$$

Function:

$$\arctan X$$

Range:

$$-1 \leq X \leq 1$$

Approximation:

$$\arctan{}^* X = \sum_{i=0}^{4} C_{2i+1} X^{2i+1}$$

$$C_1 = .9998660 \qquad C_7 = -.0851330$$
$$C_3 = -.3302995 \qquad C_9 = .0208351$$
$$C_5 = .1801410$$

Error Curve (Approximation–Function):

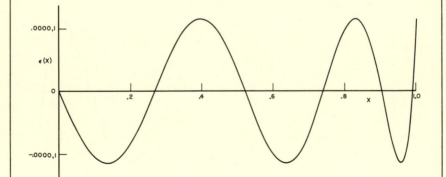

Comments: An equally good approximation over $(0 \leq X < \infty)$ is

$$\arctan{}^* X = \frac{\pi}{4} + \sum_{i=0}^{4} C_{2i+1} \left(\frac{X - 1}{X + 1} \right)^{2i+1}.$$

Function:

$$\arctan X$$

Range:

$$-1 \leq X \leq 1$$

Approximation:

$$\arctan^* X = \sum_{i=0}^{5} C_{2i+1} X^{2i+1}$$

$$C_1 = .9999,7726 \qquad C_7 = -.1164,3287$$

$$C_3 = -.3326,2347 \qquad C_9 = .0526,5332$$

$$C_5 = .1935,4346 \qquad C_{11} = -.0117,2120$$

Error Curve (Approximation–Function):

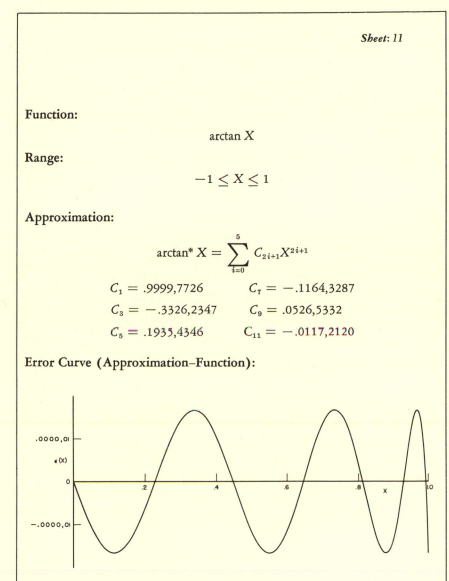

Comments: An equally good approximation over $(0 \leq X < \infty)$ is

$$\arctan^* X = \frac{\pi}{4} + \sum_{i=0}^{5} C_{2i+1}\left(\frac{X-1}{X+1}\right)^{2i+1}.$$

Function:

$$\arctan X$$

Range:

$$-1 \leq X \leq 1$$

Approximation:

$$\arctan^* X = \sum_{i=0}^{6} C_{2i+1} X^{2i+1}$$

$$C_1 = .99999,6115 \qquad C_9 = .07962,6318$$
$$C_3 = -.33317,3758 \qquad C_{11} = -.03360,6269$$
$$C_5 = .19807,8690 \qquad C_{13} = .00681,2411$$
$$C_7 = -.13233,5096$$

Error Curve (Approximation–Function):

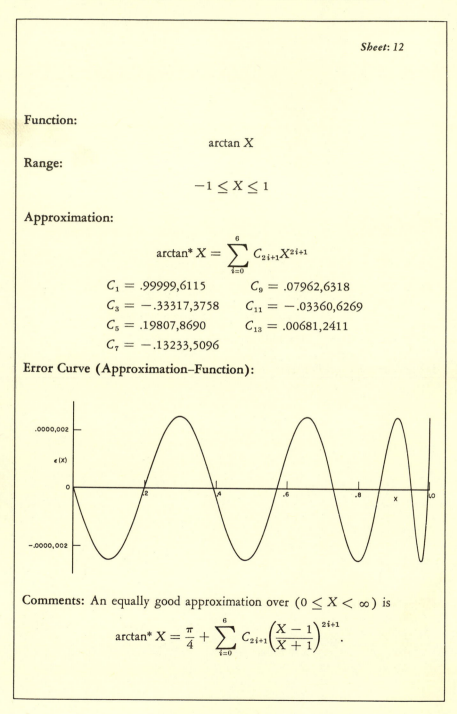

Comments: An equally good approximation over $(0 \leq X < \infty)$ is

$$\arctan^* X = \frac{\pi}{4} + \sum_{i=0}^{6} C_{2i+1} \left(\frac{X-1}{X+1} \right)^{2i+1}.$$

Function:

$$\arctan X$$

Range:

$$-1 \leq X \leq 1$$

Approximation:

$$\arctan^* X = \sum_{i=0}^{7} C_{2i+1} X^{2i+1}$$

$$C_1 = .99999,93329 \qquad C_9 = .09642,00441$$
$$C_3 = -.33329,85605 \qquad C_{11} = -.05590,98861$$
$$C_5 = .19946,53599 \qquad C_{13} = .02186,12288$$
$$C_7 = -.13908,53351 \qquad C_{15} = -.00405,40580$$

Error Curve (Approximation–Function):

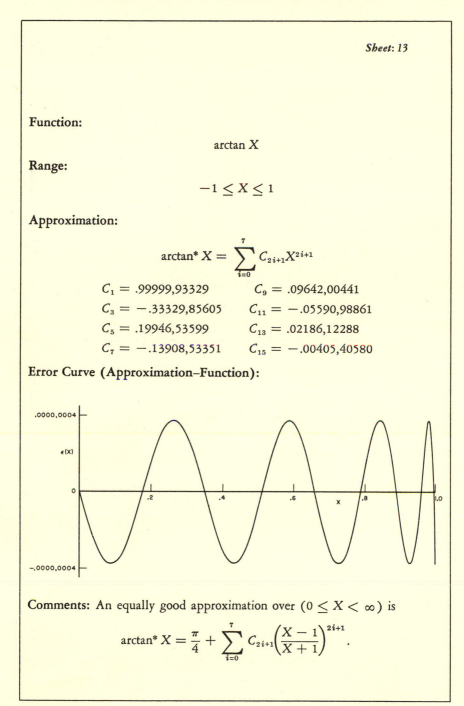

Comments: An equally good approximation over $(0 \leq X < \infty)$ is

$$\arctan^* X = \frac{\pi}{4} + \sum_{i=0}^{7} C_{2i+1}\left(\frac{X-1}{X+1}\right)^{2i+1}.$$

Function:

$$\sin \frac{\pi}{2} X$$

Range:

$$-1 \leq X \leq 1$$

Approximation:

$$\sin^* \frac{\pi}{2} X = C_1 X + C_3 X^3 + C_5 X^5$$

$$C_1 = 1.5706268$$

$$C_3 = -.6432292$$

$$C_5 = .0727102$$

Error Curve (Approximation–Function)/(Function):

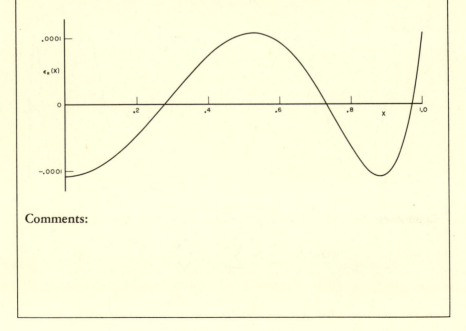

Comments:

Function:

$$\sin \frac{\pi}{2} X$$

Range:

$$-1 \leq X \leq 1$$

Approximation:

$$\sin^* \frac{\pi}{2} X = C_1 X + C_3 X^3 + C_5 X^5 + C_7 X^7$$

$$C_1 = 1.5707,94852$$

$$C_3 = -.6459,20978$$

$$C_5 = .0794,87663$$

$$C_7 = -.0043,62476$$

Error Curve (Approximation–Function)/(Function):

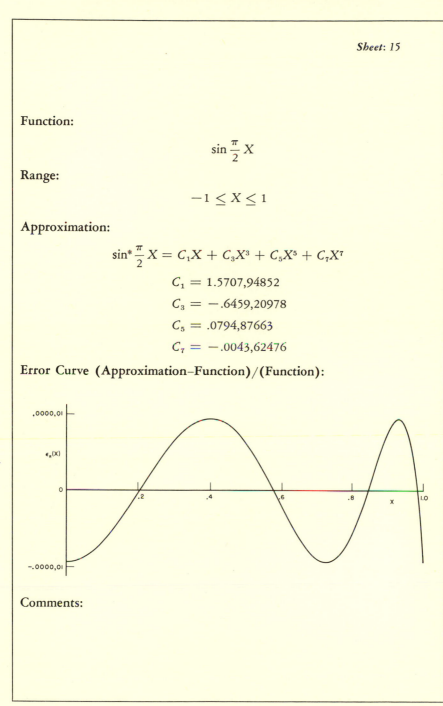

Comments:

Function:

$$\sin \frac{\pi}{2} X$$

Range:

$$-1 \leq X \leq 1$$

Approximation:

$$\sin^* \frac{\pi}{2} X = C_1 X + C_3 X^3 + C_5 X^5 + C_7 X^7 + C_9 X^9$$

$C_1 = 1.57079,631847$ $C_7 = -.00467,376557$

$C_3 = -.64596,371106$ $C_9 = .00015,148419$

$C_5 = .07968,967928$

Error Curve (Approximation–Function)/(Function):

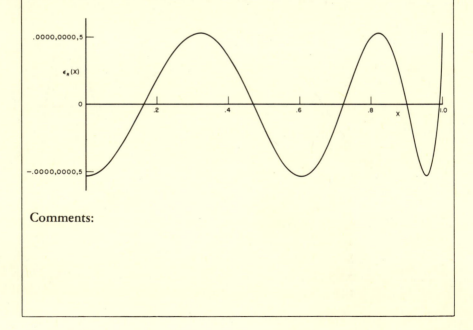

Comments:

Function:

$$10^X$$

Range:

$$0 \le X \le 1$$

Approximation:

$$(10^X)^* = [1 + a_1X + a_2X^2 + a_3X^3 + a_4X^4]^2$$

$$a_1 = 1.1499196 \qquad a_3 = .2080030$$

$$a_2 = .6774323 \qquad a_4 = .1268089$$

Error Curve (Approximation–Function)/(Function):

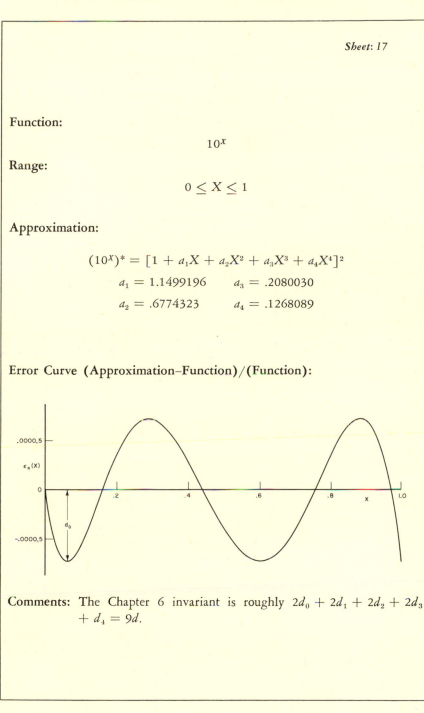

Comments: The Chapter 6 invariant is roughly $2d_0 + 2d_1 + 2d_2 + 2d_3 + d_4 = 9d$.

Function:

$$10^X$$

Range:

$$0 \leq X \leq 1$$

Approximation:

$$(10^X)^* = [1 + a_1X + a_2X^2 + a_3X^3 + a_4X^4 + a_5X^5]^2$$

$$a_1 = 1.1513,8424 \qquad a_4 = .0589,0681$$

$$a_2 = .6613,0851 \qquad a_5 = .0293,6622$$

$$a_3 = .2613,0650$$

Error Curve (Approximation–Function)/(Function):

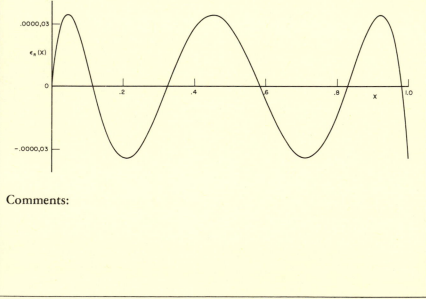

Comments:

Function:

$$10^X$$

Range:

$$0 \leq X \leq 1$$

Approximation:

$$(10^X)^* = [1 + a_1X + a_2X^2 + a_3X^3 + a_4X^4 + a_5X^5 + a_6X^6]^2$$

$$a_1 = 1.1512,87586 \qquad a_4 = .0754,67547$$

$$a_2 = .6628,43149 \qquad a_5 = .0134,20940$$

$$a_3 = .2536,03317 \qquad a_6 = .0056,54902$$

Error Curve (Approximation–Function)/(Function):

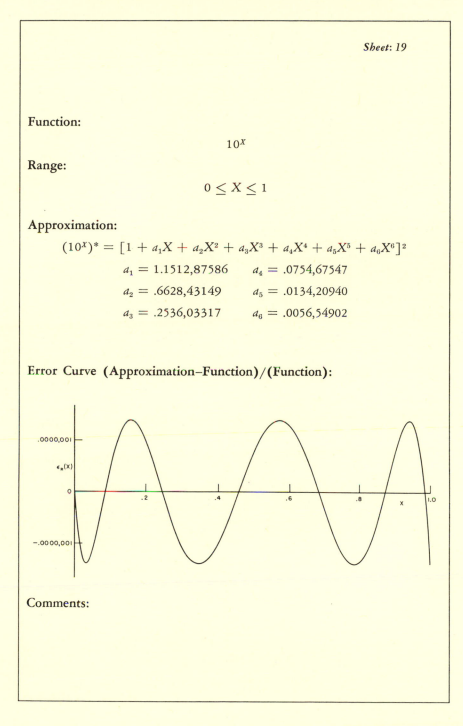

Comments:

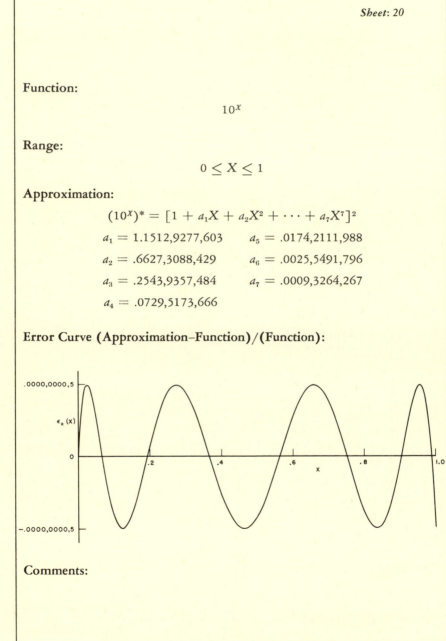

Sheet: 20

Function:

$$10^X$$

Range:

$$0 \leq X \leq 1$$

Approximation:

$$(10^X)^* = [1 + a_1X + a_2X^2 + \cdots + a_7X^7]^2$$

$a_1 = 1.1512,9277,603$ $a_5 = .0174,2111,988$

$a_2 = .6627,3088,429$ $a_6 = .0025,5491,796$

$a_3 = .2543,9357,484$ $a_7 = .0009,3264,267$

$a_4 = .0729,5173,666$

Error Curve (Approximation–Function)/(Function):

.0000,0000,5

$\epsilon_R (X)$

0

−.0000,0000,5

.2   .4   .6   .8   1.0

X

Comments:

Function:

$$W(X) = \frac{e^{-X}}{(1 + e^{-X})^2}$$

Range:

$$-\infty < X < \infty$$

Approximation:

$$W^*(X) = \frac{1}{b_0 + b_2 X^2 + b_4 X^4 + b_6 X^6}$$

$$b_0 = 3.99416 \qquad b_4 = .066512$$

$$b_2 = 1.03000 \qquad b_6 = .0048992$$

Error Curve (Approximation–Function):

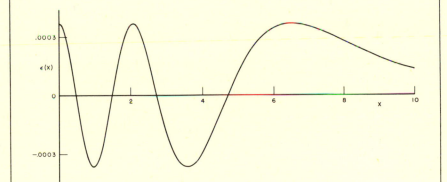

Comments: Illustrative and possibly useful in "floating decimal" computing.

Function:

$$W(X) = \frac{e^{-X}}{(1 + e^{-X})^2}$$

Range:

$$-\infty < X < \infty$$

Approximation:

$$W^*(X) = \frac{1}{b_0 + b_2 X^2 + b_4 X^4 + b_6 X^6 + b_8 X^8}$$

$$b_0 = 4.000935 \qquad b_6 = .0019864$$

$$b_2 = .994274 \qquad b_8 = .0000950$$

$$b_4 = .0874877$$

Error Curve (Approximation–Function):

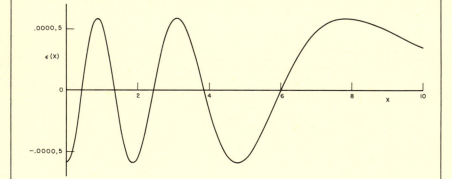

Comments: Illustrative and possibly useful in "floating decimal" computing.

Sheet: 23

Function:

$$W(X) = \frac{e^{-X}}{(1 + e^{-X})^2}$$

Range:

$$-\infty < X < \infty$$

Approximation:

$$W^*(X) = \frac{1}{b_0 + b_2 X^2 + b_4 X^4 + b_6 X^6 + b_8 X^8 + b_{10} X^{10}}$$

$b_0 = 3.9998488$ $b_6 = .00300575$

$b_2 = 1.0010596$ $b_8 = .00002974$

$b_4 = .0824114$ $b_{10} = .000001157$

Error Curve (Approximation–Function):

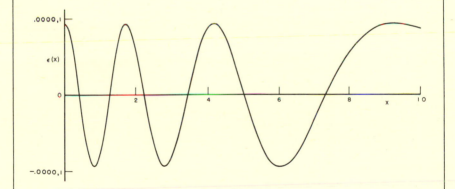

Comments: Illustrative and possibly useful in "floating decimal" computing.

Function:

$$Pk(X) = 1.72 + 42X^2 \qquad \text{over } (0, .2)$$

$$\frac{.136}{X^2} \qquad \text{over } (.2, 1)$$

Range:

$$0 \le X \le 1$$

Approximation:

$$Pk^*(X) = \frac{a_0 + a_1X + a_2X^2}{1 + b_1X + b_2X^2}$$

$$a_0 = 1.4740 \qquad b_1 = -7.5274$$

$$a_1 = -5.9044 \qquad b_2 = 17.7167$$

$$a_2 = 8.6931$$

Error Curve (Approximation–Function):

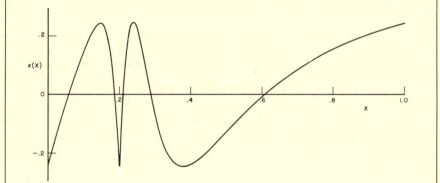

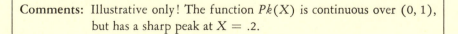

Comments: Illustrative only! The function $Pk(X)$ is continuous over $(0, 1)$, but has a sharp peak at $X = .2$.

Function:

$$Pk(X) = 1.72 + 42X^2 \qquad \text{over } (0, .2)$$

$$\frac{.136}{X^2} \qquad \text{over } (.2, 1)$$

Range:

$$0 \leq X \leq 1$$

Approximation:

$$Pk^*(X) = \frac{a_0 + a_1X + a_2X^2 + a_3X^3}{1 + b_1X + b_2X^2 + b_3X^3}$$

$a_0 = 1.83950 \qquad b_1 = -8.60126$

$a_1 = -16.45878 \qquad b_2 = 17.43641$

$a_2 = 51.11640 \qquad b_3 = 14.42411$

$a_3 = -36.09683$

Error Curve (Approximation–Function):

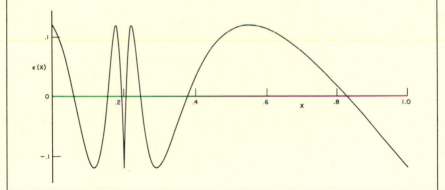

Comments: Illustrative only! The function $Pk(X)$ is continuous over $(0, 1)$, but has a sharp peak at $X = .2$.

Function:

$$Pk(X) = 1.72 + 42X^2 \qquad \text{over } (0, .2)$$

$$\frac{.136}{X^2} \qquad \text{over } (.2, 1)$$

Range:

$$0 \leq X \leq 1$$

Approximation:

$$Pk^*(X) = \frac{a_0 + a_1X + a_2X^2 + a_3X^3 + a_4X^4}{1 + b_1X + b_2X^2 + b_3X^3 + b_4X^4}$$

$a_0 = 1.653700 \qquad\qquad b_1 = -14.291266$

$a_1 = -19.425611 \qquad\; b_2 = 78.721986$

$a_2 = 78.254604 \qquad\quad b_3 = -207.451261$

$a_3 = -123.693897 \qquad b_4 = 238.403489$

$a_4 = 82.709541$

Error Curve (Approximation–Function):

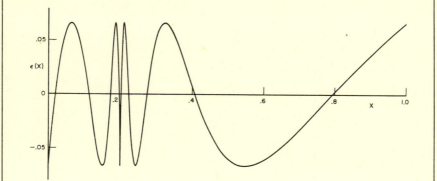
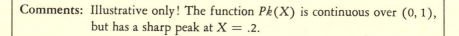

Comments: Illustrative only! The function $Pk(X)$ is continuous over $(0, 1)$, but has a sharp peak at $X = .2$.

Function:

$$E'(X) = \frac{1}{\sqrt{2\pi}} e^{-\frac{1}{2}X^2}$$

Range:

$$-\infty < X < \infty$$

Approximation:

$$\{E'(X)\}^* = \frac{1}{b_0 + b_2 X^2 + b_4 X^4 + b_6 X^6}$$

$$b_0 = 2.490895 \qquad b_4 = -.024393$$

$$b_2 = 1.466003 \qquad b_6 = .178257$$

Error Curve (Approximation–Function):

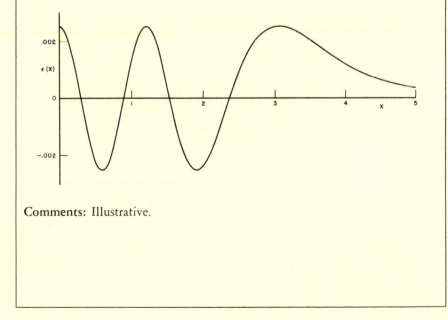

Comments: Illustrative.

Function:

$$E'(X) = \frac{1}{\sqrt{2\pi}}\, e^{-\frac{1}{2}X^2}$$

Range:

$$-\infty < X < \infty$$

Approximation:

$$\{E'(X)\}^* = \frac{1}{b_0 + b_2 X^2 + b_4 X^4 + b_6 X^6 + b_8 X^8}$$

$$b_0 = 2.511261 \qquad b_6 = -.063417$$

$$b_2 = 1.172801 \qquad b_8 = .029461$$

$$b_4 = .494618$$

Error Curve (Approximation–Function):

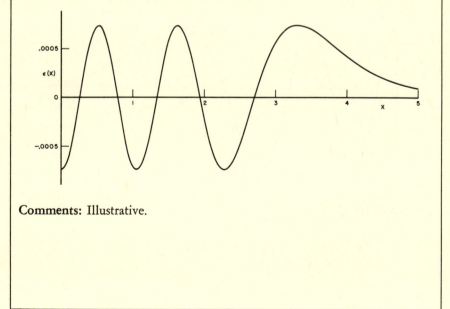

Comments: Illustrative.

Sheet: 29

Function:

$$E'(X) = \frac{1}{\sqrt{2\pi}}\, e^{-\frac{1}{2}X^2}$$

Range:

$$-\infty < X < \infty$$

Approximation:

$$\{E'(X)\}^* = \frac{1}{b_0 + b_2 X^2 + b_4 X^4 + b_6 X^6 + b_8 X^8 + b_{10} X^{10}}$$

$b_0 = 2.5052367$ $b_6 = .1306469$

$b_2 = 1.2831204$ $b_8 = -.0202490$

$b_4 = .2264718$ $b_{10} = .0039132$

Error Curve (Approximation–Function):

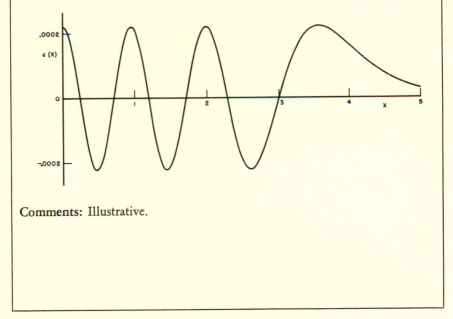

Comments: Illustrative.

Function:

$$K(n) = (n - 2n^2 - 2n^3) \ln\left(\frac{2+n}{n}\right) + \frac{(2n + 18n^2 + 16n^3 + 4n^4)}{(2+n)^2}$$

Range:

$$0 \leq n < \infty$$

Approximation:

$$\xi = \frac{n}{p+n}$$

$$K^*(n) = \frac{c_1\xi + c_2\xi^2 + c_3\xi^3}{1 + d_1\xi + d_2\xi^2 + d_3\xi^3}$$

$p = .222037$ $\quad c_1 = 1.651035$ $\quad d_1 = 12.501332$

$\qquad\qquad\qquad c_2 = 9.340220$ $\quad d_2 = -14.200407$

$\qquad\qquad\qquad c_3 = -8.325004$ $\quad d_3 = 1.699075$

Error Curve (Approximation–Function):

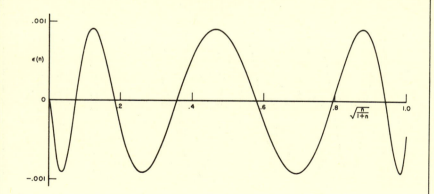

Comments: The function $K(n)$, when multiplied by the appropriate physical constant, becomes the "Total Klein-Nishina Cross Section" function. We write $K(\infty) = 8/3$.

Function:

$$\Gamma(1 + X)$$

Range:

$$0 \leq X \leq 1$$

Approximation:

$$\Gamma^*(1 + X) = 1 + a_1X + a_2X^2 + a_3X^3 + a_4X^4 + a_5X^5$$

$$a_1 = -.5748646 \qquad a_4 = .4245549$$

$$a_2 = .9512363 \qquad a_5 = -.1010678$$

$$a_3 = -.6998588$$

Error Curve (Approximation–Function):

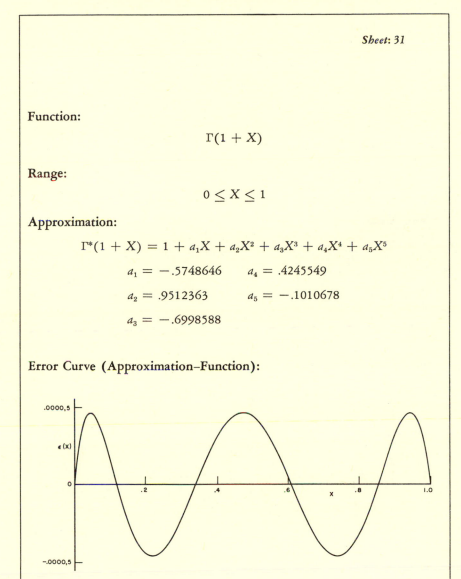

Comments:

Function:

$$\Gamma(1 + X)$$

Range:

$$0 \leq X \leq 1$$

Approximation:

$$\Gamma^*(1 + X) = 1 + a_1 X + a_2 X^2 + \cdots + a_6 X^6$$

$$a_1 = -.5766,9867 \qquad a_4 = .6739,9080$$

$$a_2 = .9778,1781 \qquad a_5 = -.3282,7930$$

$$a_3 = -.8235,6270 \qquad a_6 = .0767,3206$$

Error Curve (Approximation–Function):

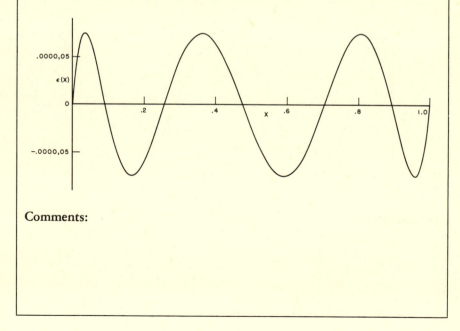

Comments:

Sheet: 33

Function:

$$\Gamma(1 + X)$$

Range:

$$0 \leq X \leq 1$$

Approximation:

$$\Gamma^*(1 + X) = 1 + a_1X + a_2X^2 + \cdots + a_7X^7$$

$$a_1 = -.5771,0166 \qquad a_5 = -.5684,7290$$

$$a_2 = .9858,5399 \qquad a_6 = .2548,2049$$

$$a_3 = -.8764,2182 \qquad a_7 = -.0514,9930$$

$$a_4 = .8328,2120$$

Error Curve (Approximation–Function):

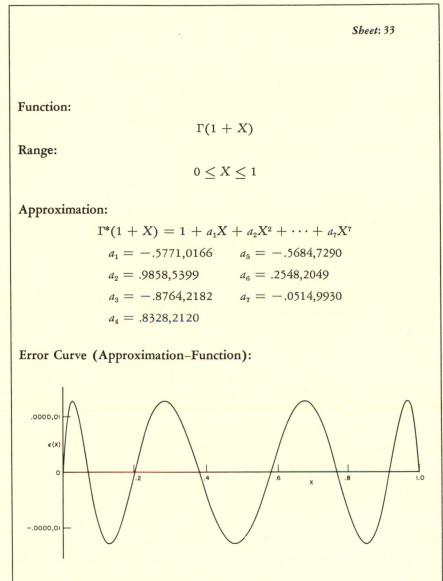

Comments:

Function:

$$\Gamma(1 + X)$$

Range:

$$0 \leq X \leq 1$$

Approximation:

$$\Gamma^*(1 + X) = 1 + a_1X + a_2X^2 + \cdots + a_8X^8$$

$$a_1 = -.5771,91652 \qquad a_5 = -.7567,04078$$

$$a_2 = .9882,05891 \qquad a_6 = .4821,99394$$

$$a_3 = -.8970,56937 \qquad a_7 = -.1935,27818$$

$$a_4 = .9182,06857 \qquad a_8 = .0358,68343$$

Error Curve (Approximation–Function):

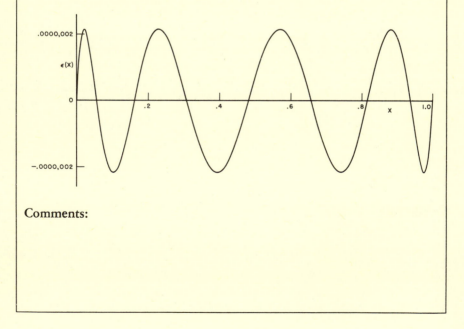

Comments:

Function:

$$\arcsin X = \frac{\pi}{2} - \sqrt{1 - X}\, \Psi(X)$$

Range:

$$0 \leq X \leq 1$$

Approximation:

$$\Psi^*(X) = a_0 + a_1 X + a_2 X^2 + a_3 X^3$$

$$a_0 = 1.5707,288 \qquad a_2 = .0742,610$$

$$a_1 = -.2121,144 \qquad a_3 = -.0187,293$$

Error Curve:

Comments:

Sheet: 36

Function:

$$\arcsin X = \frac{\pi}{2} - \sqrt{1 - X}\ \Psi(X)$$

Range:

$$0 \le X \le 1$$

Approximation:

$$\Psi^*(X) = a_0 + a_1X + a_2X^2 + a_3X^3 + a_4X^4$$

$$a_0 = 1.5707,8786 \qquad a_3 = -.0357,5663$$

$$a_1 = -.2141,2453 \qquad a_4 = .0086,4884$$

$$a_2 = .0846,6649$$

Error Curve:

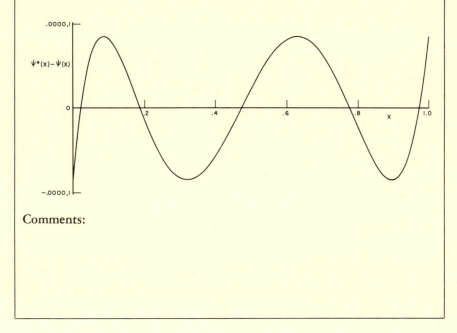

Comments:

Function:

$$\arcsin X = \frac{\pi}{2} - \sqrt{1 - X}\,\Psi(X)$$

Range:

$$0 \leq X \leq 1$$

Approximation:

$$\Psi^*(X) = a_0 + a_1 X + a_2 X^2 + a_3 X^3 + a_4 X^4 + a_5 X^5$$

$a_0 = 1.5707,95207$ $a_3 = -.0449,58884$

$a_1 = -.2145,12362$ $a_4 = .0193,49939$

$a_2 = .0878,76311$ $a_5 = -.0043,37769$

Error Curve:

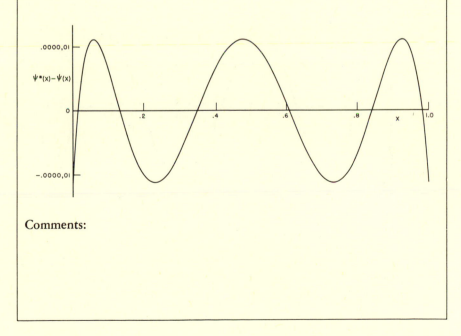

Comments:

Sheet: 38

Function:

$$\arcsin X = \frac{\pi}{2} - \sqrt{1 - X}\,\Psi(X)$$

Range:

$$0 \le X \le 1$$

Approximation:

$$\Psi^*(X) = a_0 + a_1 X + a_2 X^2 + \cdots + a_6 X^6$$

$a_0 = 1.5707,9617,28$ $a_4 = .0268,9994,82$

$a_1 = -.2145,8526,47$ $a_5 = -.0111,4622,94$

$a_2 = .0887,5562,86$ $a_6 = .0022,9596,48$

$a_3 = -.0488,0250,43$

Error Curve:

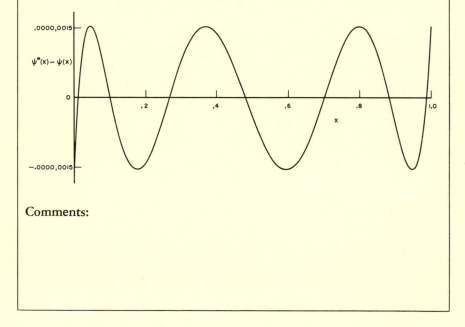

Comments:

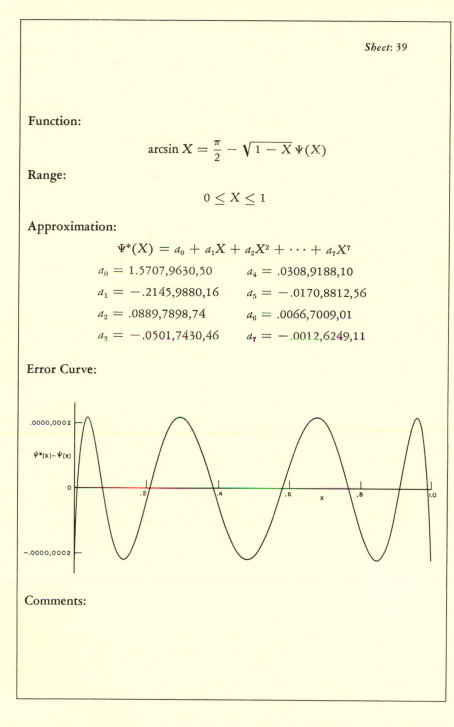

Function:

$$\arcsin X = \frac{\pi}{2} - \sqrt{1 - X}\,\Psi(X)$$

Range:

$$0 \le X \le 1$$

Approximation:

$$\Psi^*(X) = a_0 + a_1 X + a_2 X^2 + \cdots + a_7 X^7$$

$a_0 = 1.5707,9630,50$ $a_4 = .0308,9188,10$

$a_1 = -.2145,9880,16$ $a_5 = -.0170,8812,56$

$a_2 = .0889,7898,74$ $a_6 = .0066,7009,01$

$a_3 = -.0501,7430,46$ $a_7 = -.0012,6249,11$

Error Curve:

Comments:

Function:

$$\log_2 X$$

Range:

$$\frac{1}{\sqrt{2}} \le X \le \sqrt{2}$$

Approximation:

$$\log_2^* X = c_1\left(\frac{X-1}{X+1}\right) + c_3\left(\frac{X-1}{X+1}\right)^3$$

$$c_1 = 2.8852,2873$$

$$c_3 = .9835,2829$$

Error Curve (Approximation–Function):

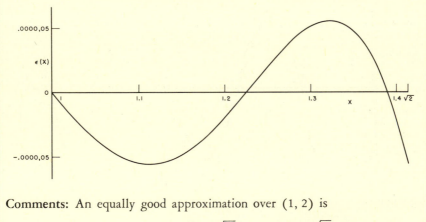

Comments: An equally good approximation over $(1, 2)$ is

$$\log_2^* X = \frac{1}{2} + c_1\left(\frac{X-\sqrt{2}}{X+\sqrt{2}}\right) + c_3\left(\frac{X-\sqrt{2}}{X+\sqrt{2}}\right)^3 .$$

Function:

$$\log_2 X$$

Range:

$$\frac{1}{\sqrt{2}} \leq X \leq \sqrt{2}$$

Approximation:

$$\log_2^* X = c_1\left(\frac{X-1}{X+1}\right) + c_3\left(\frac{X-1}{X+1}\right)^3 + c_5\left(\frac{X-1}{X+1}\right)^5$$

$$c_1 = 2.8853,9129,03$$
$$c_3 = .9614,7063,23$$
$$c_5 = .5989,7864,96$$

Error Curve (Approximation–Function):

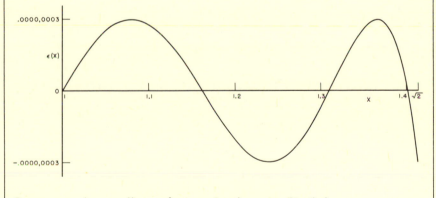

Comments: An equally good approximation over $(1, 2)$ is

$$\log_2^* X = \frac{1}{2} + c_1\left(\frac{X-\sqrt{2}}{X+\sqrt{2}}\right) + c_3\left(\frac{X-\sqrt{2}}{X+\sqrt{2}}\right)^3 + c_5\left(\frac{X-\sqrt{2}}{X+\sqrt{2}}\right)^5 .$$

Sheet: 42

Function:

$$\log_2 X$$

Range:

$$\frac{1}{\sqrt{2}} \le X \le \sqrt{2}$$

Approximation:

$$\log_2^* X = c_1\left(\frac{X-1}{X+1}\right) + c_3\left(\frac{X-1}{X+1}\right)^3 + c_5\left(\frac{X-1}{X+1}\right)^5 + c_7\left(\frac{X-1}{X+1}\right)^7$$

$$c_1 = 2.8853,9007,2738 \qquad c_5 = .5765,8434,2056$$

$$c_3 = .9618,0076,2286 \qquad c_7 = .4342,5975,1292$$

Error Curve (Approximation–Function):

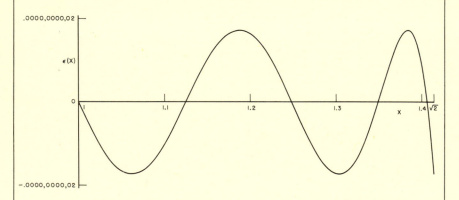

Comments: An equally good approximation over $(1, 2)$ is

$$\log_2^* X = \frac{1}{2} + c_1\left(\frac{X-\sqrt{2}}{X+\sqrt{2}}\right) + c_3\left(\frac{X-\sqrt{2}}{X+\sqrt{2}}\right)^3 + \cdots + c_7\left(\frac{X-\sqrt{2}}{X+\sqrt{2}}\right)^7.$$

Function:

$$\Phi(X) = \frac{2}{\sqrt{\pi}} \int_0^X e^{-t^2}\, dt$$

Range:

$$0 \leq X < \infty$$

Approximation:

$$\eta = \frac{1}{1 + pX}$$

$$\Phi^*(X) = 1 - (a_1\eta + a_2\eta^2 + a_3\eta^3)\Phi'(X)$$

$$p = .47047 \qquad a_1 = .3084{,}284$$

$$a_2 = -.0849{,}713$$

$$a_3 = .6627{,}698$$

Error Curve (Approximation–Function):

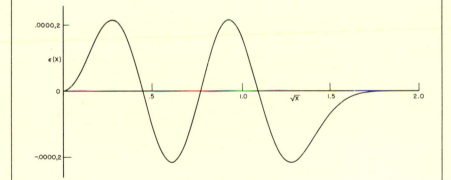

Comments: Especially useful when $\Phi'(X)$ must also be computed.

Function:

$$\Phi(X) = \frac{2}{\sqrt{\pi}} \int_0^X e^{-t^2}\, dt$$

Range:

$$0 \leq X < \infty$$

Approximation:

$$\eta = \frac{1}{1 + pX}$$

$$\Phi^*(X) = 1 - (a_1\eta + a_2\eta^2 + a_3\eta^3 + a_4\eta^4)\Phi'(X)$$

$$p = .381965 \qquad a_1 = .1277,1538$$

$$a_2 = .5410,7939$$

$$a_3 = -.5385,9539$$

$$a_4 = .7560,2755$$

Error Curve (Approximation–Function):

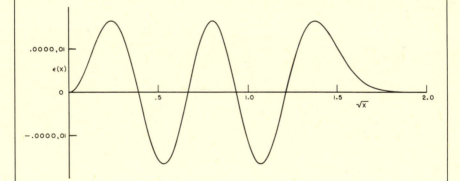

Comments: Especially useful when $\Phi'(X)$ must also be computed.

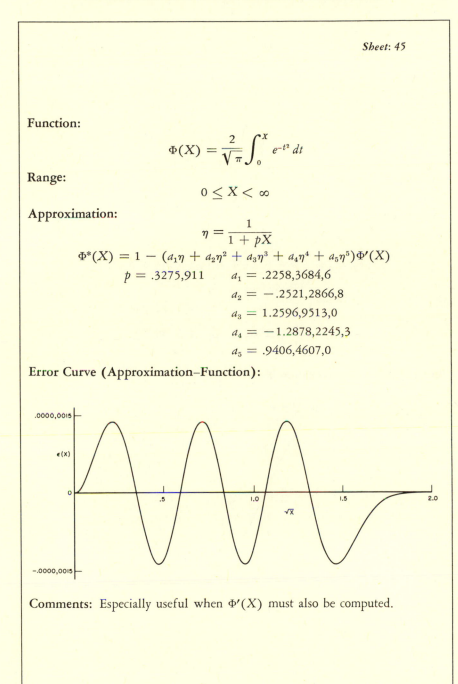

Function:

$$\Phi(X) = \frac{2}{\sqrt{\pi}} \int_0^X e^{-t^2}\, dt$$

Range:

$$0 \le X < \infty$$

Approximation:

$$\eta = \frac{1}{1 + pX}$$

$$\Phi^*(X) = 1 - (a_1\eta + a_2\eta^2 + a_3\eta^3 + a_4\eta^4 + a_5\eta^5)\Phi'(X)$$

$$p = .3275,911 \qquad a_1 = .2258,3684,6$$
$$a_2 = -.2521,2866,8$$
$$a_3 = 1.2596,9513,0$$
$$a_4 = -1.2878,2245,3$$
$$a_5 = .9406,4607,0$$

Error Curve (Approximation–Function):

Comments: Especially useful when $\Phi'(X)$ must also be computed.

Function:

$$K(k) = \int_0^{\pi/2} \frac{d\varphi}{\sqrt{1 - k^2 \sin^2 \varphi}}$$

Range:

$$0 \leq k < 1$$

Approximation:

$$\eta = 1 - k^2$$

$$K^*(k) = \{a_0 + a_1\eta + a_2\eta^2\} + \{b_0 + b_1\eta + b_2\eta^2\} \ln\frac{1}{\eta}$$

$$a_0 = 1.3862,944 \qquad b_0 = .5$$

$$a_1 = .1119,723 \qquad b_1 = .1213,478$$

$$a_2 = .0725,296 \qquad b_2 = .0288,729$$

Error Curve (Approximation–Function):

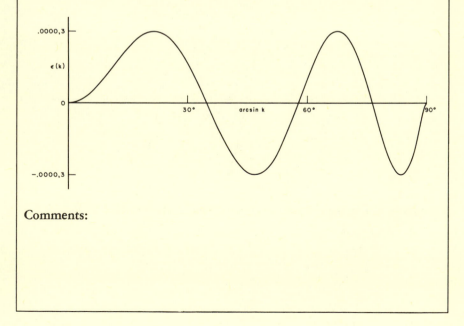

Comments:

Function:

$$K(k) = \int_0^{\pi/2} \frac{d\varphi}{\sqrt{1 - k^2 \sin^2 \varphi}}$$

Range:

$$0 \leq k < 1$$

Approximation:

$$\eta = 1 - k^2$$

$$K^*(k) = \{a_0 + a_1\eta + a_2\eta^2 + a_3\eta^3\} + \{b_0 + b_1\eta + b_2\eta^2 + b_3\eta^3\} \ln\frac{1}{\eta}$$

$$a_0 = 1.3862,9436,1 \qquad b_0 = .5$$
$$a_1 = .0979,3289,1 \qquad b_1 = .1247,5074,2$$
$$a_2 = .0545,4440,9 \qquad b_2 = .0601,1851,9$$
$$a_3 = .0320,2466,6 \qquad b_3 = .0109,4491,2$$

Error Curve (Approximation–Function):

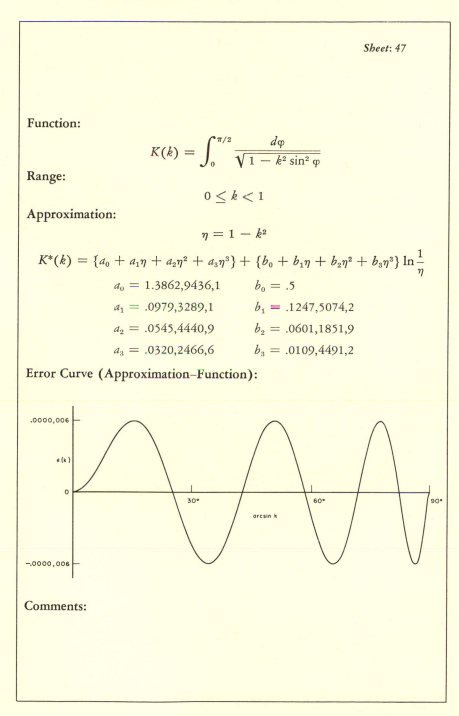

Comments:

Function:

$$K(k) = \int_0^{\pi/2} \frac{d\varphi}{\sqrt{1 - k^2 \sin^2 \varphi}}$$

Range:

$$0 \leq k < 1$$

Approximation:

$$\eta = 1 - k^2$$

$$K^*(k) = \{a_0 + a_1\eta + \cdots + a_4\eta^4\} + \{b_0 + b_1\eta + \cdots + b_4\eta^4\} \ln \frac{1}{\eta}$$

$a_0 = 1.3862,9436,112$ $b_0 = .5$

$a_1 = .0966,6344,259$ $b_1 = .1249,8593,597$

$a_2 = .0359,0092,383$ $b_2 = .0688,0248,576$

$a_3 = .0374,2563,713$ $b_3 = .0332,8355,346$

$a_4 = .0145,1196,212$ $b_4 = .0044,1787,012$

Error Curve (Approximation–Function):

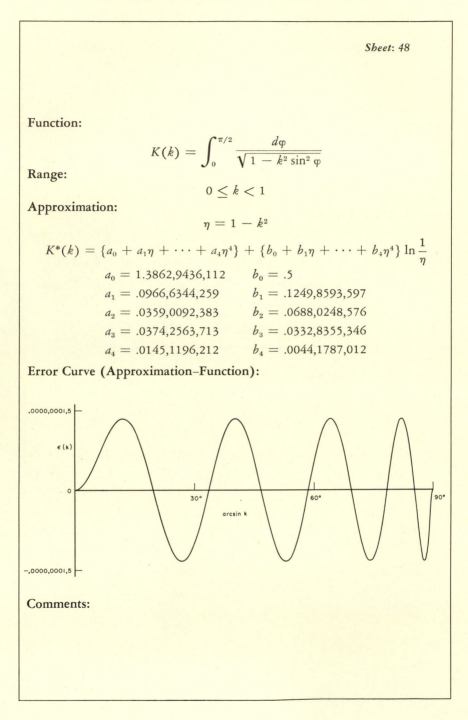

Comments:

Function:

$$E(k) = \int_0^{\pi/2} \sqrt{1 - k^2 \sin^2 \varphi}\, d\varphi$$

Range:

$$0 \leq k < 1$$

Approximation:

$$\eta = 1 - k^2$$

$$E^*(k) = \{1 + a_1\eta + a_2\eta^2\} + \{b_1\eta + b_2\eta^2\} \ln\frac{1}{\eta}$$

$$a_1 = .4630,151 \qquad b_1 = .2452,727$$

$$a_2 = .1077,812 \qquad b_2 = .0412,496$$

Error Curve (Approximation–Function):

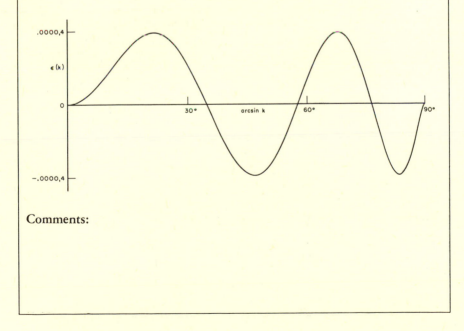

Comments:

Function:

$$E(k) = \int_0^{\pi/2} \sqrt{1 - k^2 \sin^2 \varphi} \, d\varphi$$

Range:

$$0 \leq k < 1$$

Approximation:

$$\eta = 1 - k^2$$

$$E^*(k) = \{1 + a_1\eta + a_2\eta^2 + a_3\eta^3\} + \{b_1\eta + b_2\eta^2 + b_3\eta^3\} \ln\frac{1}{\eta}$$

$$a_1 = .4447,9204,0 \qquad b_1 = .2496,9794,9$$

$$a_2 = .0850,9919,3 \qquad b_2 = .0815,0224,0$$

$$a_3 = .0409,0509,4 \qquad b_3 = .0138,2999,0$$

Error Curve (Approximation–Function):

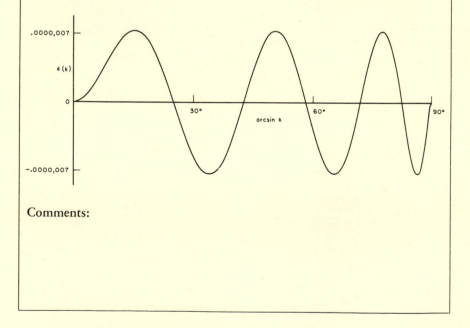

Comments:

Function:

$$E(k) = \int_0^{\pi/2} \sqrt{1 - k^2 \sin^2 \varphi} \, d\varphi$$

Range:

$$0 \leq k < 1$$

Approximation:

$$\eta = 1 - k^2$$

$$E^*(k) = \{1 + a_1\eta + \cdots + a_4\eta^4\} + \{b_1\eta + \cdots + b_4\eta^4\} \ln\frac{1}{\eta}$$

$$a_1 = .4432,5141,463 \qquad b_1 = .2499,8368,310$$
$$a_2 = .0626,0601,220 \qquad b_2 = .0920,0180,037$$
$$a_3 = .0475,7383,546 \qquad b_3 = .0406,9697,526$$
$$a_4 = .0173,6506,451 \qquad b_4 = .0052,6449,639$$

Error Curve (Approximation–Function):

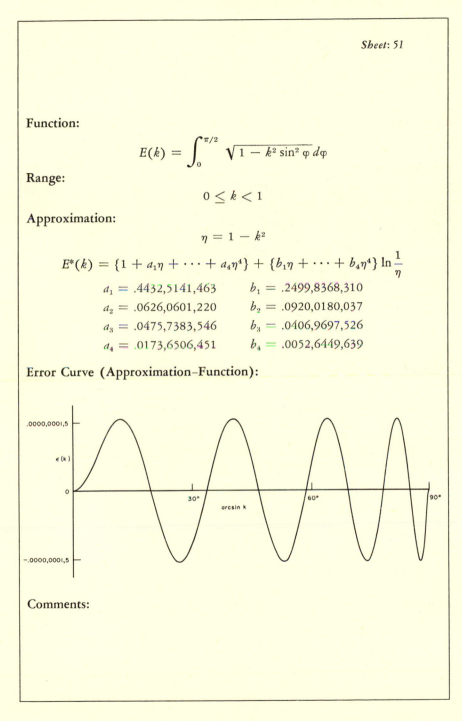

Comments:

Function:

$$\ln(1 + X)$$

Range:

$$0 \leq X \leq 1$$

Approximation:

$$\ln^*(1 + X) = a_1 X + a_2 X^2 + a_3 X^3 + a_4 X^4$$

$$a_1 = .9974{,}442 \qquad a_3 = .2256{,}685$$

$$a_2 = -.4712{,}839 \qquad a_4 = -.0587{,}527$$

Error Curve (Approximation–Function):

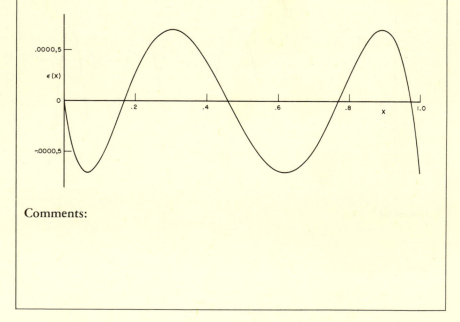

Comments:

Sheet: 53

Function:

$$\ln (1 + X)$$

Range:

$$0 \leq X \leq 1$$

Approximation:

$$\ln^* (1 + X) = a_1X + a_2X^2 + a_3X^3 + a_4X^4 + a_5X^5$$

$$a_1 = .9994,9556 \qquad a_4 = -.1360,6275$$

$$a_2 = -.4919,0896 \qquad a_5 = .0321,5845$$

$$a_3 = .2894,7478$$

Error Curve (Approximation–Function):

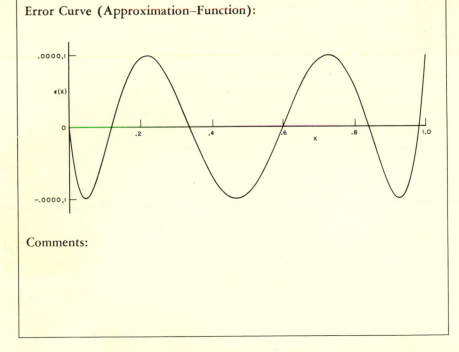

Comments:

Function:

$$\ln (1 + X)$$

Range:

$$0 \leq X \leq 1$$

Approximation:

$$\ln^* (1 + X) = a_1 X + a_2 X^2 + \cdots + a_6 X^6$$

$a_1 = .9999,0167$ $a_4 = -.1937,6149$

$a_2 = -.4978,7544$ $a_5 = .0855,6927$

$a_3 = .3176,5005$ $a_6 = -.0183,3831$

Error Curve (Approximation–Function):

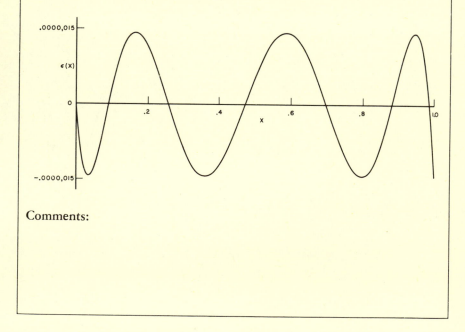

Comments:

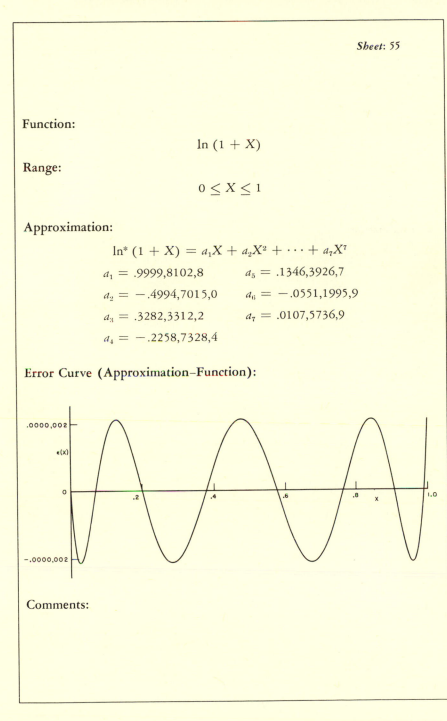

Function:

$$\ln (1 + X)$$

Range:

$$0 \leq X \leq 1$$

Approximation:

$$\ln^* (1 + X) = a_1 X + a_2 X^2 + \cdots + a_7 X^7$$

$a_1 = .9999,8102,8$ $a_5 = .1346,3926,7$

$a_2 = -.4994,7015,0$ $a_6 = -.0551,1995,9$

$a_3 = .3282,3312,2$ $a_7 = .0107,5736,9$

$a_4 = -.2258,7328,4$

Error Curve (Approximation–Function):

Comments:

Function:

$$\ln (1 + X)$$

Range:

$$0 \leq X \leq 1$$

Approximation:

$$\ln^* (1 + X) = a_1X + a_2X^2 + \cdots + a_8X^8$$

$a_1 = .9999,9642,39$ $a_5 = .1676,5407,11$

$a_2 = -.4998,7412,38$ $a_6 = -.0953,2938,97$

$a_3 = .3317,9902,58$ $a_7 = .0360,8849,37$

$a_4 = -.2407,3380,84$ $a_8 = -.0064,5354,42$

Error Curve (Approximation–Function):

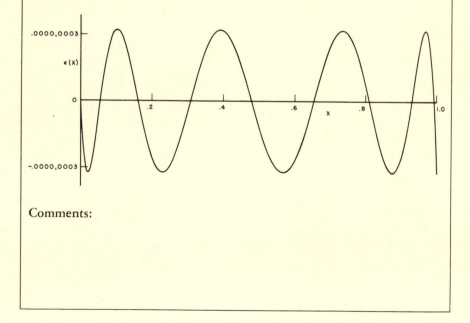

Comments:

Function:

$$e^{-X}$$

Range:

$$0 \leq X < \infty$$

Approximation:

$$\{e^{-X}\}^* = \frac{1}{[1 + a_1X + a_2X^2 + a_3X^3]^4}$$

$$a_1 = .2507,213$$

$$a_2 = .0292,732$$

$$a_3 = .0038,278$$

Error Curve (Approximation–Function):

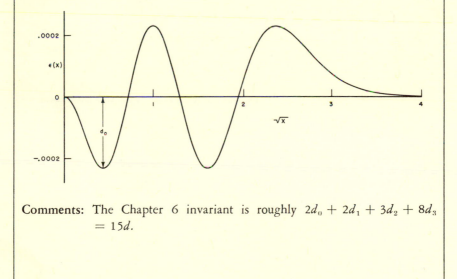

Comments: The Chapter 6 invariant is roughly $2d_0 + 2d_1 + 3d_2 + 8d_3$
$= 15d$.

Function:

$$e^{-X}$$

Range:

$$0 \leq X < \infty$$

Approximation:

$$\{e^{-X}\}^* = \frac{1}{[1 + a_1X + a_2X^2 + a_3X^3 + a_4X^4]^4}$$

$$a_1 = .2499,1035 \qquad a_3 = .0022,7723$$

$$a_2 = .0315,8565 \qquad a_4 = .0002,6695$$

Error Curve (Approximation–Function):

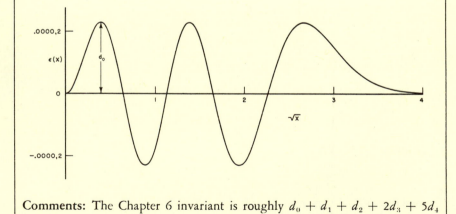

Comments: The Chapter 6 invariant is roughly $d_0 + d_1 + d_2 + 2d_3 + 5d_4$
$= 10d.$

Function:

$$e^{-X}$$

Range:

$$0 \leq X < \infty$$

Approximation:

$$\{e^{-X}\}^* = \frac{1}{[1 + a_1X + a_2X^2 + a_3X^3 + a_4X^4 + a_5X^5]^4}$$

$$a_1 = .2500,1093,6 \qquad a_4 = .0001,2799,2$$

$$a_2 = .0311,9805,6 \qquad a_5 = .0000,1487,6$$

$$a_3 = .0026,7325,5$$

Error Curve (Approximation–Function):

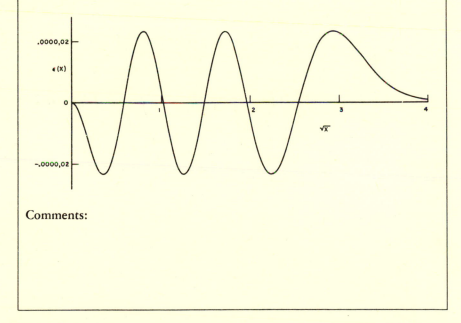

Comments:

Sheet: 60

Function:

$$e^{-X}$$

Range:

$$0 \leq X < \infty$$

Approximation:

$$\{e^{-X}\}^* = \frac{1}{[1 + a_1X + a_2X^2 + a_3X^3 + a_4X^4 + a_5X^5 + a_6X^6]^4}$$

$a_1 = .2499,9868,42$ $a_4 = .0001,7156,20$

$a_2 = .0312,5758,32$ $a_5 = .0000,0543,02$

$a_3 = .0025,9137,12$ $a_6 = .0000,0069,06$

Error Curve (Approximation–Function):

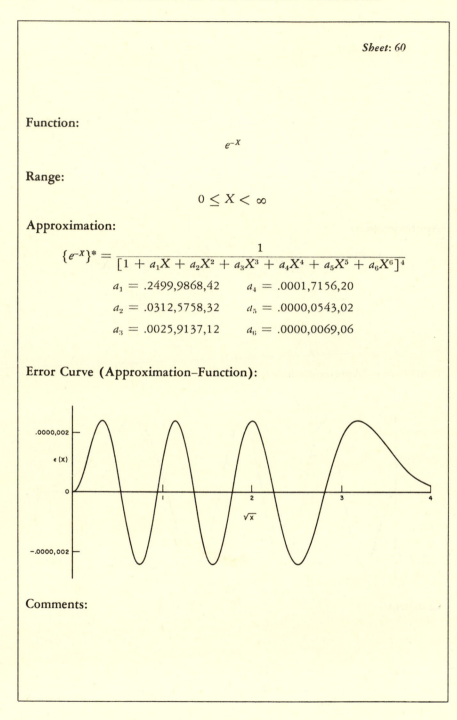

Comments:

Function:

$$\Phi(X) = \frac{2}{\sqrt{\pi}} \int_0^X e^{-t^2}\, dt$$

Range:

$$0 \leq X < \infty$$

Approximation:

$$\Phi^*(X) = 1 - \frac{1}{[1 + a_1 X + a_2 X^2 + a_3 X^3 + a_4 X^4]^4}$$

$$a_1 = .278393 \qquad a_3 = .000972$$

$$a_2 = .230389 \qquad a_4 = .078108$$

Error Curve (Approximation–Function):

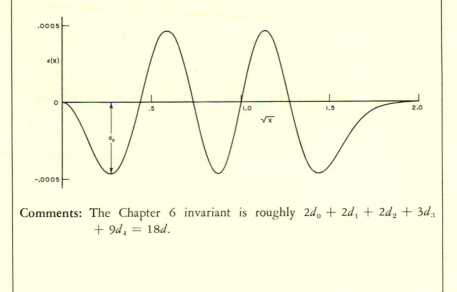

Comments: The Chapter 6 invariant is roughly $2d_0 + 2d_1 + 2d_2 + 3d_3 + 9d_4 = 18d$.

Sheet: 62

Function:

$$\Phi(X) = \frac{2}{\sqrt{\pi}} \int_0^X e^{-t^2} \, dt$$

Range:

$$0 \leq X < \infty$$

Approximation:

$$\Phi^*(X) = 1 - \frac{1}{[1 + a_1X + a_2X^2 + a_3X^3 + a_4X^4 + a_5X^5]^8}$$

$a_1 = .1411,2821$ $a_4 = -.0003,9446$

$a_2 = .0886,4027$ $a_5 = .0032,8975$

$a_3 = .0274,3349$

Error Curve (Approximation–Function):

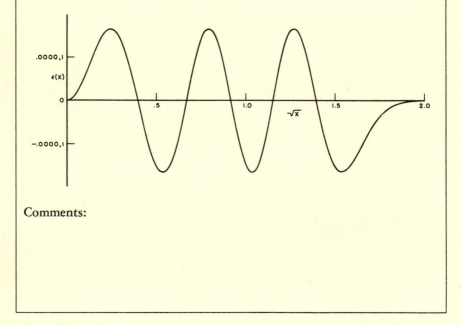

Comments:

Function:

$$\Phi(X) = \frac{2}{\sqrt{\pi}} \int_0^X e^{-t^2}\, dt$$

Range:

$$0 \leq X < \infty$$

Approximation:

$$\Phi^*(X) = 1 - \frac{1}{[1 + a_1X + a_2X^2 + a_3X^3 + a_4X^4 + a_5X^5 + a_6X^6]^{16}}$$

$a_1 = .0705,2307,84$ $a_4 = .0001,5201,43$

$a_2 = .0422,8201,23$ $a_5 = .0002,7656,72$

$a_3 = .0092,7052,72$ $a_6 = .0000,4306,38$

Error Curve (Approximation–Function):

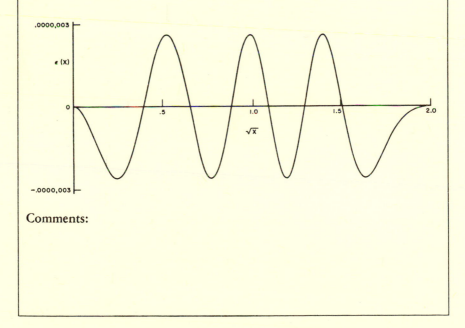

Comments:

Function:

$$-Ei(-X) = \int_X^\infty \frac{e^{-t}}{t}\, dt$$

Range:

$$1 \le X < \infty$$

Approximation:

$$-Ei^*(-X) = \frac{e^{-X}}{X}\left\{\frac{a_0 + a_1 X + X^2}{b_0 + b_1 X + X^2}\right\}$$

$$a_0 = .250621 \qquad b_0 = 1.681534$$

$$a_1 = 2.334733 \qquad b_1 = 3.330657$$

Error Curve (Approximation–Function)/(Function):

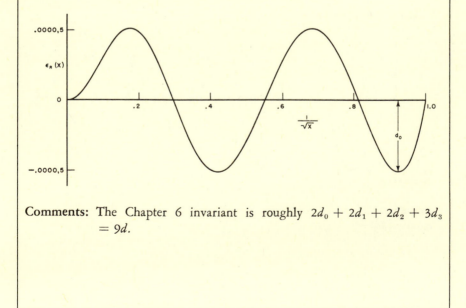

Comments: The Chapter 6 invariant is roughly $2d_0 + 2d_1 + 2d_2 + 3d_3 = 9d$.

Function:

$$-Ei(-X) = \int_X^\infty \frac{e^{-t}\,dt}{t}$$

Range:

$$1 \leq X < \infty$$

Approximation:

$$-Ei^*(-X) = \frac{e^{-X}}{X}\left\{\frac{a_0 + a_1 X + a_2 X^2 + X^3}{b_0 + b_1 X + b_2 X^2 + X^3}\right\}$$

$a_0 = .2372,9050 \qquad b_0 = 2.4766,3307$

$a_1 = 4.5307,9235 \qquad b_1 = 8.6660,1262$

$a_2 = 5.1266,9020 \qquad b_2 = 6.1265,2717$

Error Curve (Approximation–Function)/(Function):

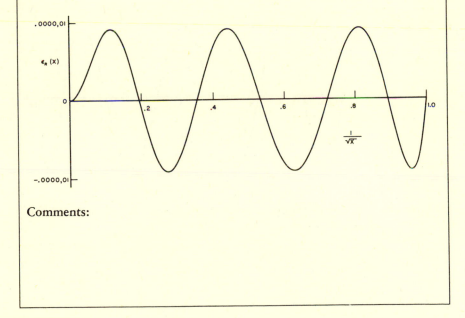

Comments:

Function:

$$-Ei(-X) = \int_{X}^{\infty} \frac{e^{-t}}{t}\, dt$$

Range:

$$1 \le X < \infty$$

Approximation:

$$-Ei^{*}(-X) = \frac{e^{-X}}{X}\left\{\frac{a_0 + a_1 X + a_2 X^2 + a_3 X^3 + X^4}{b_0 + b_1 X + b_2 X^2 + b_3 X^3 + X^4}\right\}$$

$a_0 = .2677,7373,43$ $b_0 = 3.9584,9692,28$

$a_1 = 8.6347,6089,25$ $b_1 = 21.0996,5308,27$

$a_2 = 18.0590,1697,30$ $b_2 = 25.6329,5614,86$

$a_3 = 8.5733,2874,01$ $b_3 = 9.5733,2234,54$

Error Curve (Approximation–Function)/(Function):

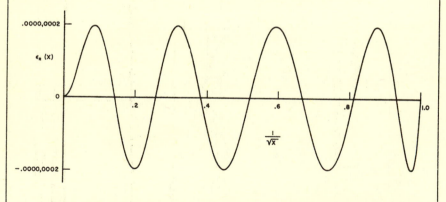

Comments: The approximation is based on our own $18D$ estimates of the function $-Xe^{X}Ei(-X)$.

Function:

$$q = \frac{1}{\sqrt{2\pi}} \int_{X(q)}^{\infty} e^{-\frac{1}{2}t^2}\, dt$$

Range:

$$0 < q \leq .5$$

Approximation:

$$\eta = \sqrt{\ln \frac{1}{q^2}}$$

$$X^*(q) = \eta - \left\{ \frac{a_0 + a_1 \eta}{1 + b_1 \eta + b_2 \eta^2} \right\}$$

$$a_0 = 2.30753 \qquad b_1 = .99229$$

$$a_1 = .27061 \qquad b_2 = .04481$$

Error Curve (Approximation–Function):

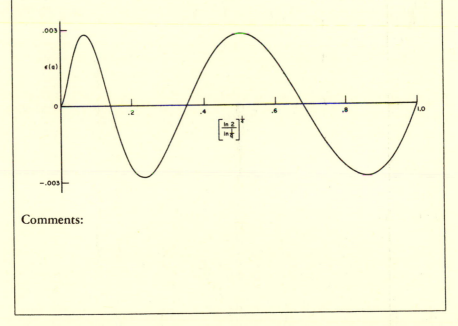

Comments:

Function:

$$q = \frac{1}{\sqrt{2\pi}} \int_{X(q)}^{\infty} e^{-\frac{1}{2}t^2} \, dt$$

Range:

$$0 < q \leq .5$$

Approximation:

$$\eta = \sqrt{\ln \frac{1}{q^2}}$$

$$X^*(q) = \eta - \left\{ \frac{a_0 + a_1\eta + a_2\eta^2}{1 + b_1\eta + b_2\eta^2 + b_3\eta^3} \right\}$$

$$a_0 = 2.515517 \qquad b_1 = 1.432788$$

$$a_1 = .802853 \qquad b_2 = .189269$$

$$a_2 = .010328 \qquad b_3 = .001308$$

Error Curve (Approximation–Function):

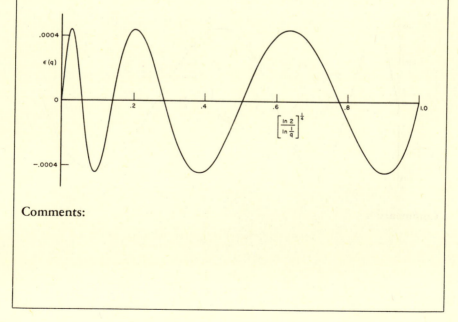

Comments:

Sheet: *69*

Function:

$$W(Z) = \int_0^\infty \frac{e^{-uZ}}{K_1^2(u) + \pi^2 I_1^2(u)} \frac{du}{u}$$

Range:

$$0 \leq Z < \infty$$

Approximation:

$$W^*(Z) = \frac{1 + a_1 Z}{2 + b_1 Z + b_2 Z^2 + b_3 Z^3}$$

$$a_1 = .058689 \qquad b_1 = 1.624877$$

$$b_2 = .440731$$

$$b_3 = .084386$$

Error Curve (Approximation–Function):

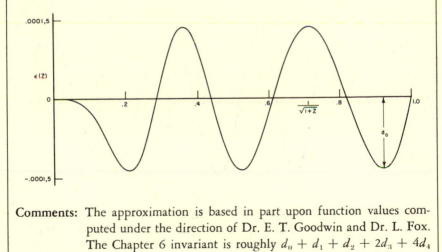

Comments: The approximation is based in part upon function values computed under the direction of Dr. E. T. Goodwin and Dr. L. Fox. The Chapter 6 invariant is roughly $d_0 + d_1 + d_2 + 2d_3 + 4d_4 = 9d$.

Sheet: 70

Function:

$$W(Z) = \int_0^\infty \frac{e^{-uZ}}{K_1^2(u) + \pi^2 I_1^2(u)} \frac{du}{u}$$

Range:

$$0 \le Z < \infty$$

Approximation:

$$W^*(Z) = \frac{1 + a_1 Z + a_2 Z^2}{2 + b_1 Z + b_2 Z^2 + b_3 Z^3 + b_4 Z^4}$$

$a_1 = .0490,1768$ $b_1 = 1.5975,1756$

$a_2 = .0074,9940$ $b_2 = .4659,6355$

$b_3 = .0732,2506$

$b_4 = .0070,0350$

Error Curve (Approximation–Function):

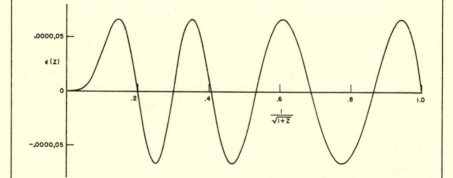

Comments: Dr. E. T. Goodwin and his associates in the Mathematics Division of the National Physical Laboratory, Teddington, Middlesex, England, very graciously computed for us the 9D table of $W(Z)$ upon which this approximation is based.

Function:

$$P(X) = \int_X^\infty \frac{\sin(t - X)\,dt}{t}$$

Range:

$$1 \le X < \infty$$

Approximation:

$$P^*(X) = \frac{1}{X}\left\{\frac{a_0 + a_2 X^2 + X^4}{b_0 + b_2 X^2 + X^4}\right\}$$

$$a_0 = 2.463936 \qquad b_0 = 7.157433$$

$$a_2 = 7.241163 \qquad b_2 = 9.068580$$

Error Curve (Approximation–Function):

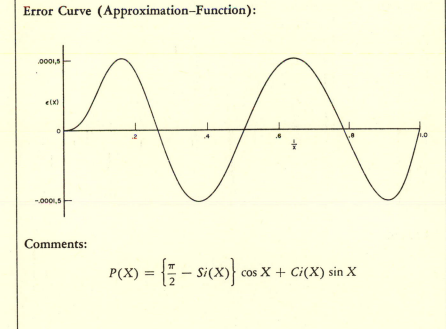

Comments:

$$P(X) = \left\{\frac{\pi}{2} - Si(X)\right\}\cos X + Ci(X)\sin X$$

Function:

$$P(X) = \int_X^\infty \frac{\sin (t - X)\, dt}{t}$$

Range:

$$1 \le X < \infty$$

Approximation:

$$P^*(X) = \frac{1}{X} \left\{ \frac{a_0 + a_2 X^2 + a_4 X^4 + X^6}{b_0 + b_2 X^2 + b_4 X^4 + X^6} \right\}$$

$a_0 = 8.493336$ $b_0 = 30.038277$

$a_2 = 47.411538$ $b_2 = 70.376496$

$a_4 = 19.394119$ $b_4 = 21.361055$

Error Curve (Approximation–Function):

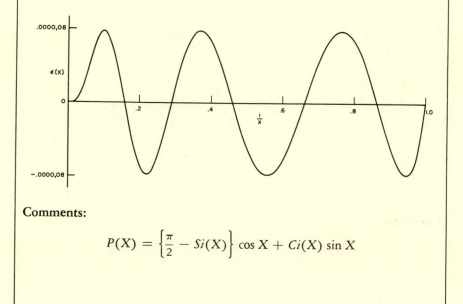

Comments:

$$P(X) = \left\{ \frac{\pi}{2} - Si(X) \right\} \cos X + Ci(X) \sin X$$

Function:

$$P(X) = \int_X^\infty \frac{\sin{(t - X)}\, dt}{t}$$

Range:

$$1 \leq X < \infty$$

Approximation:

$$P^*(X) = \frac{1}{X}\left\{\frac{a_0 + a_2 X^2 + a_4 X^4 + a_6 X^6 + X^8}{b_0 + b_2 X^2 + b_4 X^4 + b_6 X^6 + X^8}\right\}$$

$a_0 = 38.102495 \qquad b_0 = 157.105423$

$a_2 = 335.677320 \qquad b_2 = 570.236280$

$a_4 = 265.187033 \qquad b_4 = 322.624911$

$a_6 = 38.027264 \qquad b_6 = 40.021433$

Error Curve (Approximation–Function):

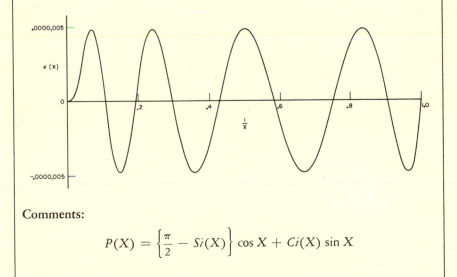

Comments:

$$P(X) = \left\{\frac{\pi}{2} - Si(X)\right\}\cos X + Ci(X)\sin X$$

Function:

$$Q(X) = \int_X^\infty \frac{\cos(t - X)\, dt}{t}$$

Range:

$$1 \leq X < \infty$$

Approximation:

$$Q^*(X) = \frac{1}{X^2}\left\{\frac{a_0 + a_2 X^2 + X^4}{b_0 + b_2 X^2 + X^4}\right\}$$

$$a_0 = 1.564072 \qquad b_0 = 15.723606$$

$$a_2 = 7.547478 \qquad b_2 = 12.723684$$

Error Curve (Approximation–Function):

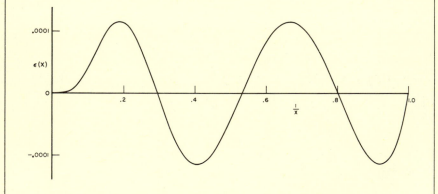

Comments:

$$Q(X) = \left\{\frac{\pi}{2} - Si(X)\right\} \sin X - Ci(X) \cos X$$

Function:

$$Q(X) = \int_X^\infty \frac{\cos(t - X)\,dt}{t}$$

Range:

$$1 \leq X < \infty$$

Approximation:

$$Q^*(X) = \frac{1}{X^2}\left\{\frac{a_0 + a_2 X^2 + a_4 X^4 + X^6}{b_0 + b_2 X^2 + b_4 X^4 + X^6}\right\}$$

$$a_0 = 5.089504 \qquad b_0 = 76.707878$$

$$a_2 = 49.719775 \qquad b_2 = 119.918932$$

$$a_4 = 21.383724 \qquad b_4 = 27.177958$$

Error Curve (Approximation–Function):

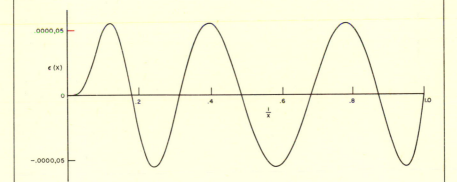

Comments:

$$Q(X) = \left\{\frac{\pi}{2} - Si(X)\right\}\sin X - Ci(X)\cos X$$

Function:

$$Q(X) = \int_{X}^{\infty} \frac{\cos (t - X) \, dt}{t}$$

Range:

$$1 \leq X < \infty$$

Approximation:

$$Q^*(X) = \frac{1}{X^2} \left\{ \frac{a_0 + a_2 X^2 + a_4 X^4 + a_6 X^6 + X^8}{b_0 + b_2 X^2 + b_4 X^4 + b_6 X^6 + X^8} \right\}$$

$$a_0 = 21.821899 \qquad b_0 = 449.690326$$

$$a_2 = 352.018498 \qquad b_2 = 1114.978885$$

$$a_4 = 302.757865 \qquad b_4 = 482.485984$$

$$a_6 = 42.242855 \qquad b_6 = 48.196927$$

Error Curve (Approximation–Function):

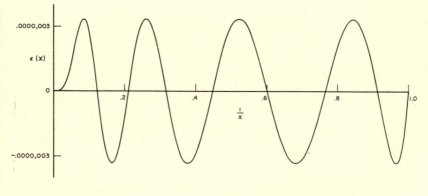

Comments:

$$Q(X) = \left\{ \frac{\pi}{2} - Si(X) \right\} \sin X - Ci(X) \cos X$$

REFERENCES FOR PART II

1. British Association for the Advancement of Science, *Mathematical Tables*, Vol. 1, Cambridge University Press, London, 1946.
2. BRUHNS, C., *A New Manual of Logarithms*, The Charles T. Powner Company, Chicago, 1942.
3. CALLET, FRANCOIS, AND ERNEST C. BOWER, "Table of Sines and Cosines for Decimal Parts of the Circle," unpublished but available on punched cards at The RAND Corporation, Santa Monica, California. See *Mathematical Tables and Other Aids to Computation* (MTAC), Vol. 3, No. 26, April, 1949, pp. 425–426.
4. KELLEY, TRUMAN LEE, *The Kelley Statistical Tables*, The Macmillan Company, New York, 1938.
5. LEGENDRE, ADRIEN-MARIE, AND KARL PEARSON, *Tables of the Complete and Incomplete Elliptic Integrals*, Cambridge University Press, London, 1934.
6. National Bureau of Standards: all pertinent books of mathematical tables.
7. National Research Council, *Mathematical Tables and other Aids to Computation* (MTAC). Approximations of ours are published from time to time in this journal, beginning with the January, 1953, issue.
8. SPENCELEY, G. W. AND R. M., *Smithsonian Elliptic Functions Tables* (Smithsonian Miscellaneous Collections, Vol. 109), Publication 3863, Smithsonian Institution, Washington, D.C., 1947. We referred to photostats of the original manuscript obtained from the Brown University library.
9. WARD, G. N., "The Approximate External and Internal Flow Past a Quasi-cylindrical Tube Moving at Supersonic Speeds," *The Quarterly Journal of Mechanics and Applied Mathematics*, Vol. 1, 1948, pp. 225–245. The function $W(Z)$ of Sheets 69 and 70 is discussed in this paper.

Selected RAND Books

Bellman, Richard. *Adaptive Control Processes: A Guided Tour*. Princeton, N.J.: Princeton University Press, 1961.

Bellman, Richard. *Dynamic Programming*. Princeton, N.J.: Princeton University Press, 1957.

Bellman, Richard, and Kenneth L. Cooke. *Differential-Difference Equations*. New York: Academic Press, 1963.

Bellman, Richard, and Stuart E. Dreyfus. *Applied Dynamic Programming*. Princeton, N.J.: Princeton University Press, 1962.

Bellman, Richard E., Harriet H. Kagiwada, Robert E. Kalaba, and Marcia C. Prestrud. *Invariant Imbedding and Time-Dependent Transport Processes*, Modern Analytic and Computational Methods in Science and Mathematics, Vol. 2. New York: American Elsevier Publishing Company, Inc., 1964.

Bellman, Richard E., and Robert E. Kalaba. *Quasilinearization and Nonlinear Boundary-Value Problems*, Modern Analytic and Computational Methods in Science and Mathematics, Vol. 3, New York: American Elsevier Publishing Company, Inc., 1965.

Bellman, Richard E., Robert E. Kalaba, and Marcia C. Prestrud. *Invariant Imbedding and Radiative Transfer in Slabs of Finite Thickness*, Modern Analytic and Computational Methods in Science and Mathematics, Vol. 1. New York: American Elsevier Publishing Company, Inc., 1963.

Buchheim, Robert W., and the Staff of The RAND Corporation. *New Space Handbook: Astronautics and Its Applications*. New York: Vintage Books, A Division of Random House, Inc., 1963.

Dantzig, G. B. *Linear Programming and Extensions*. Princeton, N.J.: Princeton University Press, 1963.

Dresher, Melvin. *Games of Strategy: Theory and Applications*. Englewood Cliffs, N.J.: Prentice-Hall, Inc., 1961.

Edelin, Dominic G. B. *The Structure of Field Space: An Axiomatic Formulation of Field Physics*. Berkeley and Los Angeles: University of California Press, 1962.

Ford, L. R., Jr., and D. R. Fulkerson. *Flows in Networks*. Princeton, N.J.: Princeton University Press, 1962.

Harris, Theodore E. *The Theory of Branching Processes*. Berlin, Germany: Springer-Verlag, 1963; Englewood Cliffs, N.J.: Prentice-Hall, Inc., 1964.

Hitch, Charles J., and Roland McKean. *The Economics of Defense in the Nuclear Age*. Cambridge, Mass.: Harvard University Press, 1960.

Quade, Edward S. (ed.). *Analysis for Military Decisions*. Chicago: Rand McNally & Company; Amsterdam: North-Holland Publishing Company, 1964.

Williams, J. D. *The Compleat Strategyst: Being a Primer on the Theory of Games of Strategy*. New York: McGraw-Hill Book Company, Inc., 1954.